Taghi Erani, a Polymath in Interwar Berlin

Younes Jalali

Taghi Erani, a Polymath in Interwar Berlin

Fundamental Science, Psychology, Orientalism, and Political Philosophy

Younes Jalali
Saint-Nom-la-Bretèche, Yvelines
France

ISBN 978-3-319-97836-9   ISBN 978-3-319-97837-6   (eBook)
https://doi.org/10.1007/978-3-319-97837-6

Library of Congress Control Number: 2018950509

This Palgrave Macmillan imprint is published by the registered company Springer Nature
Switzerland AG
The registered company address is: Gewerbestrasse 11, 6330 Cham, Switzerland

For Forough and Asadollah

PREFACE

This is the story of someone who came of age in 1920s Berlin, a time and place with no shortage of shorthand in popular culture and mythology—the capital of eccentricity; a city that doubled its population overnight, four million strong, incorporating the settlements and hamlets on its periphery by administrative ordinance; with an institute dedicated to the study of sex, another to alchemy of the mind, another demystifying life as making and breaking of chemical bonds; Austro-Hungarian refugees, white Russians, maimed veterans, cafes, newsstands, jazz clubs, a "Republic without Republicans";[1] the right wanting their emperor, the left communism, very little in-between; Wedding and Neukölln preparing for class warfare; barricades of Berlin in run-down neighborhoods in the run-up to May Day; social democrats in bed with right-wing parties; the left seeking salvation in Bolshevism; Comintern fronting for the Bolsheviks, run by unscrupulous characters, Berlin its plotting factory; genius in science, evil genius in politics; audacity in art, literature, theater, architecture, Dadaism, psychoanalysis, experimentation on all fronts. There is perhaps no better way to catch a glimpse of this universe than by following the trail of an inquisitive mind and daring soul exposing its grittiness and naked truths.

Our protagonist was a high-ranking civil servant and intellectual whose worldview was forged in 1920s Berlin, at the height of Weimar culture, and his political bent with a strong accent on social radicalism and anti-war pacifism led to his downfall in 1930s Persia—*The Erani Affair*—in the aftermath of stints in the Third Reich which produced eyewitness accounts as riveting as Joseph Roth's *feuilletons* from Nazi Berlin.

His story reveals not only the genetic makeup and anatomy of Pahlavi during its founding and formative stage—a dynasty of two kings spurred by great powers, superseding the lethargic but indigenous Qajar dynasty (seven kings)—a disposition that led to its collapse in the 1979 revolution, it also reveals the minute mechanics of geopolitical posturing in the late interwar period, on the eve of outbreak of World War II, themes with a certain resonance to contemporary history and its interwar-like undertones.

A journey in modern political and intellectual history intersecting key events and personalities encountered by the protagonist in high science, psychology, orientalism, international communism, and dynastic politics based on archives of Weimar, Pahlavi, and Comintern, as well as eyewitness accounts of key actors and family archives of the protagonist.

Saint-Nom-la-Bretèche, France   Younes Jalali
May 2018

NOTE

1. *The Ailing Empire, Germany from Bismarck to Hitler*, Sebastian Haffner, 1989, p. 159.

ACKNOWLEDGEMENTS

During 1986–1987, I developed correspondence with five personalities who had collaborated with or known Dr. Taghi Erani—Mr. S.M.A. Jamalzadeh (Geneva), Dr. Mehdi Azar (Tokyo), Mr. Bozorg Alavi (Berlin), Mr. Ayromlou (Cologne), and Dr. Anvar Khame'i (Tehran). I am grateful for their patience and care in answering my questions.

Contact with the above eminences was made possible by Dr. Fereydoon Keschavarz, Mr. Mohammad Assemi, Prof. Homayoun Katouzian, Mr. Ali Zarrabi, Mr. Ahmad Bashiri, Mr. Mohammad Golbon, and Mr. Iraj Afshar.

The relatives of Dr. Taghi Erani have wished to remain anonymous, but their contribution to this work, notably providing access to family archives, was most valuable in reconstructing his life story. However, all representations of his life contained in this work are the responsibility of the author and do not necessarily reflect the views of the Erani family. The Erani family affirm that Dr. Erani's prolific life was one of scientific, intellectual, and cultural contributions and his political life was within the bounds of legality as guaranteed by the Iranian constitution of 1906.

Ms. Azadeh Akhlaghi shared her insights about the demise of Dr. Taghi Erani, as captured in her artistic recreation in the collection "By an Eyewitness."

Mrs. Anne Namini-Saunders did a meticulous job to free my writing of its convolutedness and grammatical inexactitudes. I would also like to thank Ms. Megan Laddusaw, Palgrave Acquisitions Editor for History, and her team particularly Ms. Christine Pardue for the superb job in accompanying the work throughout the production process.

Mr. Chris-Julian Fruhner examined Taghi Erani's doctoral dissertation and produced a summary in English. Dr. Stefan Reich-Albrecht oversaw the work. Mrs. Ines Albrecht coordinated the project. I would also like to thank the Albrecht family for having shared their unique insights and perspective on this work with the publisher.

Archivists from former East and West Germany were most resourceful in helping me find original materials from the Weimar era pertaining to Taghi Erani, as were bibliographers from McGill, UCLA, ELO (Paris), Majles library (Tehran), Brown (private collection of Prof. Archibald), Hoover, Chicago, and other institutes, 1980s vintage.

I benefited enormously from prior archival findings and interpretations of scholars who had worked on Taghi Erani—Prof. Ahmad Mahrad (Hannover), Mr. Bagher Momeni (Berlin), Mr. Hossein Boroujerdi (Tehran), Dr. Cosroe Chaqueri (Paris), and Dr. Hamid Ahmadi (Berlin). I had correspondence with Prof. Mahrad and Dr. Chaqueri.

Mr. Parviz Shokat provided an account of his visit of 2008 with the youngest sister of Taghi Erani. Dr. Yusef Jalali provided photographs of burial site of Taghi Erani, as well as access to the all-important *Parvandeh*. Ms. Azadeh Moselm brought back a rare copy of Taghi Erani's textbook on chemistry from Tehran (Majles library). Mr. Salehpour provided translations of select archival materials from German to Persian.

Professor Colin Atkinson, FRS, Imperial College, was my most loyal reader and relentless critic; he drew my attention to Orwell's *Homage to Catalonia*, as parallel to the Erani story, how the Comintern was fixated on eliminating forces of independent left.

Professor Suzanne Marchand, LSU, doyen of modern German intellectual history and orientalism, meticulously reviewed my early drafts and provided much guidance on how to make it more substantive. Professor Stephanie Cronin of Oxford University, foremost scholar on modern Iranian history, notably the Reza Shah period, generously agreed to have the author propose her name as reviewer to the publisher. Professor Michael Wertheimer, University of Colorado, reviewed my rendering of European psychology of interwar years and provided several corrections and clarifications. Professor Christoph Werner, Marburg, Iranian Studies, provided feedback on my draft.

Ms. Leslie Gardner, Artellus, provided insights on the publishing world; Mr. Hooman Majd, author, made this connection possible. A number of friends read fragments of the work and provided helpful feedback—Edgar Dias, Allyson Gajraj, Songming Huang, Shiva Pakdel, Jurgen Conrad, Takeo Fujihara, Zhang Lingyu, Charlie Cosad.

Professor Franklin M. Orr, Jr., Stanford, Undersecretary of Energy in the Obama administration, permitted me to use his weighty name in my literary bio, on account of our teaching collaboration in years bygone.

Professor Noam Chomsky reviewed early drafts and provided much encouragement.

The Jalali family provided enormous emotional and intellectual support over long years—my dear wife Rezika and children—Mauni, Shireen, Jasmin; as well my siblings—Mahnaz, Maryam, Ebi, Yusef. This work would not have been possible without them.

CONTENTS

Note on Translation

Translation from Persian to English of passages in Dr. Erani's works is done by the present author; brackets indicate insertions by the translator; the rest are present in the original, including text in the parentheses.

CHAPTER 1

The Arrest (1937)

Taghi Erani was arrested by Reza Shah's police on May 8, 1937, at his residence in Tehran. It was a Saturday, the start of the week, Friday being the normal day off. He lived with his mother and youngest sister, who was seventeen years old at the time. They had a house-servant (*gomaushte*) who had been working for them over twenty years and was in his early forties. The other sisters of Taghi Erani, in their mid-twenties, had been married and lived in the south, the oil province, as their spouses worked in that region. One worked for the Anglo-Persian Oil Company (APOC) in Masjed-Soleiman, 800-km from the capital, site of a giant oilfield. The other worked for the Trans-Iranian Railroad (TIRR) in Ahvaz, 850-km from Tehran, and 150-km to the oil terminal in Abadan, site of a giant refinery with access to the Persian Gulf. Erani's father had passed away a few months earlier, in Tabriz.

Tabriz is where Taghi Erani was born, in 1902. He had lived there until the age of ten; then, with his mother and two sisters, had moved to Tehran. This was before the outbreak of the Great War (1914–1918), which spilled into western Persia, notably Azerbaijan Province, Tabriz being its capital. Tehran is where Erani's third sister was born, after the war, his father straddling between Tabriz and Tehran. In 1922, at the age of twenty, having completed a two-year post-secondary program at the Medical College in Tehran, he left for Berlin. He stayed there six and half years, at the height of the Weimar era, while Persia was going

© The Author(s) 2019
Y. Jalali, *Taghi Erani, a Polymath in Interwar Berlin*,
https://doi.org/10.1007/978-3-319-97837-6_1

through dynastic convulsions with the demise of Qajar and the rise of Pahlavi. During this period, Erani had notable achievements in fundamental science (chemistry and physics), psychology, and orientalism; along with firsthand encounters with a number of great names in these domains and in Weimar political life. He left Berlin in December 1928, under precarious circumstances, as he had dabbled in anti-Pahlavi activism while in Berlin. Even though his activism was covert, suspicions were growing about him with the threat of expulsion. (Germany maintained cooperative relations with government of Persia in view of escalading trade. This led to the crackdown on expatriate political activists on German soil.) Erani returned to Persia in January 1929, as a way of shedding suspicions, and for five years he led a seemingly quiet life which focused on his professional career. Eight years on from his return, having earned a certain notoriety in societal circles, including appointment as Vice-Minister of Industry, he now found himself in police custody. He had come full circle.

His mother's home, where he lived, was in *Moghareb-ol-Saltaneh*, in the west side of Tehran, two kilometers from the more modest *Sheikh Hadi*, the neighborhood they had lived in when they first arrived in Tehran in 1912. The police arrived at his residence, accompanied by a man who identified Taghi Erani as the one who was acquainted with "the doctor." The witness claimed that he had come to this residence with the doctor, on their way to some other destination. He claimed that "the doctor" and Erani had spoken only in German in his presence, evidently to disguise the content of their conversation. He did not know the name of Taghi Erani but was able to identify him by sight. Taghi Erani asserted that neither did he know this individual, nor did he have a relationship of a political character with the doctor; the two were simply friends from their Berlin days. Nevertheless, he was detained by the police.

The same day, the same man accompanied the police to the house of a Berlin-educated medical doctor. He knew the name and the residence of this doctor, which was at *Sheikh Hadi*. The doctor's house, where his medical clinic was set up, had allegedly been the hub of a clandestine group. The professional character of the address, with streams of visitors, was presumed to give it a natural cover. The informer indicated he would come there, give the codename given to him by a certain *Tambourak* ("little lute"), pick up a package, usually money, and get on with his business. He would take people across the border to Russia or bring individuals into Persia. He was a smuggler, about fifty years old and of a working-class background. He was a machinist by trade (*sohan-kar*) and dabbled in theater for working-class folks. The doctor denied knowing this man and found the allegation of subversive activity absurd; nevertheless, the police detained the doctor as well.

The following day, the same man took the police to a busy intersection, *Gomrok* (customs depot), and after sometime he noticed his usual contact but hesitated to approach him. The contact, however, noticed the man, approached him, and was immediately arrested by the police. The informer did not know the name of this contact, and only knew him by his alias, a female name, *Fattane*. The latter operated a house which served as a "motel," facilitating the clandestine group's movements between the provinces and the capital. Real motels were not safe and were under police surveillance. According to a second version, the informer led the police not to the busy intersection but to this house-motel, and this was where the police arrested *Fattane*.

Subsequent to these three arrests, forty other arrests were made, nearly all within a week, the last one on June 3. The detainees represented an incongruous group: a Vice-Minister, university students and two professors; high school teachers and administrators; doctors, lawyers, clerks, and government administrators; a merchant, a novelist who earned a living as a teacher, and a mix of laboring types—print-setters (two), tailors (two), a textile worker, a cobbler, a baker, a mechanic, a locomotive conductor, a gardener, and several with unsteady jobs.

It was difficult to imagine how such disparate group could constitute a political party, association or sect. Only a few had had prior offenses, and they came from all over the country; some from the north (Gilan province, Rasht and Port Enzeli, and Mazandaran province, Gorgan), some from the south (Khuzestan, the oil province, Ahvaz and Abadan), some from Qazvin (ancient capital, 150-km west of Tehran), and many from Tehran. The youngest was not yet twenty, the oldest in his fifties; some utterly poor, some affluent, the majority were middle-class individuals.

The detainees were subjected to severe interrogation and intimidation methods at the police headquarters in the center of Tehran, at the infamous *Shahr'bani*. Swedish padlocks were heavily used, as the Swedes had founded Persia's modern police force and inter-city security, the Gendarmerie (c. 1910). There were seven interrogators, as far as records reveal.[1]

Taghi Erani and the doctor denied any wrongdoing or engagement in covert activity. Nevertheless, news broke rapidly about the arrest of "Dr. Erani and his group." Taghi Erani was a public figure, a high-ranking government official known to the court, the cabinet of ministers, the German ambassador, and the leading intellectuals.

Taghi Erani had been invited by the Prime Minister, Mahmoud Djam, to the king's birthday ceremony at the *Golistan* Palace, only weeks before the arrest (March 15); and earlier, to the joint wedding of the king's two daughters (March 4). He was a known name among the German expatriates in Tehran. He was the hierarchical superior of Dr. Schtrüng, the Head of the School of Industry, formerly the German School, which fell under the jurisdiction of the Ministry of Industry. The German ambassador, Wipert von Blücher, an orientalist by training, personally knew Taghi Erani and would invite him to social functions thrown by the German legation. Erani's association with Friedrich Rosen in Berlin, Foreign Minister under Weimar and an avid orientalist, was not lost on the German ambassador.

Taghi Erani had published a number of books in several scientific domains. Some of his works were used as textbooks by high school and university students. Erani had taught in some of these educational establishments. He had been the editor of a scientific-philosophical journal to which several prominent figures, including a *Majles* deputy, had made regular contributions. He was a household name amongst the intellectuals who charted new paths in theater, literature, music, and painting. He was known among leading members of the medical and legal profession. He was a close acquaintance of the personal tutor of the king's male offspring (the *Shapurs*); the brother and half brothers of the crown prince (who had just returned from Rosey). He was Vice-Minister of Industry, possibly poised for a higher office, court invitations being a likely signal, coming in the wake of Schacht's visit to Persia (November 1936). Schacht was the Third Reich economic strategist.

And now, out of the blue, had come the arrest. Was it a mistake? Was it genuine? Was it a plot? If so, by whom, and for what purpose? Or, was it a hodgepodge—a security operation, with ulterior motives and agendas, fraught with errors; errors exorbitantly costly for the victims, their families, and in some sense, for the nation at large. And the lingering questions—was it the king or his entourage? Was it for domestic or foreign consumption? Was it the work of governments and politicians or operatives and intelligence services? Was it a fallout of the Bajanov and Agabekov defections? If so, why so belatedly?[2]

For those affected, it was a shock and a grave tragedy; for others mind-boggling, perhaps a message directed at them via rebound and reverberation; "*to tell the door so the wall will hear,*" as the saying goes in Persian. The Persia of the Reza Shah era was not foreign to the sight of falling giants and elites. There had been a string of murders and deaths of illustrious personalities.

Prince Firouz was one of the king's earliest advocates and his Finance Minister. Firouz had recklessly hinted (to relatives of the decapitated Colonel Pasyan, a nationalist figure, an episode which implicated Reza Khan as an accomplice to Qavam-ol-Saltaneh, principal nemesis of Pasyan) that the British, meaning Lord Curzon, had earmarked him to succeed the irresolute Qajar monarch, Ahmad Shah. Whereas General Ironside had opted for Reza Khan; and being the man on the ground in Persia, the commander of the British forces, Norperforce, General Ironside had had his way. Firouz was picked up on charges of financial impropriety, of which he was guilty, for he was a compulsive overspender and was suitably eliminated.[3]

Teymurtash, the all-powerful Court Minister, and mastermind of the Reza Shah system, of aristocratic extraction, also fell and was eliminated. Teymurtash was educated in Tsarist Russia. He was a statesman of the late Qajar era and a polyglot; a man with a panoramic vision of his time. He stunned diplomats with his brilliance, and some were utterly convinced of his genius. He fell after the British divulged to the king that he was a Soviet mole with designs on the monarchy. This happened while Teymurtash was pursuing a hardnosed agenda in negotiating the oil concession renewal.[4]

Davar, Minister of Justice and later Finance, took his own life by opium overdose, understanding full well that his days were numbered. He knew his fate in view of the king's body language and utterings. He was the last of the triumvirate (Firouz, Teymurtash, Davar) that had enabled Reza Khan's power-grab in the guise of legality, virtually an impossible task given his brutish ways. Even while the fifth *Majles* was voting for the abolition of Qajar and the sanctification of Pahlavi, December

1925, he made sure that there are plenty of Cossacks and gunshots fired in the vicinity of the *Majles* so nobody got cold feet and there was no last-minute change of heart: yet five of about one hundred deputies voted against the bill and several abstained, by their absence.

There were many more high-flying victims—Sardar Asad (Bakhtiyari baron, Minister of War), Modarres (influential deputy and cleric), Mohammad-Vali Asadi (administrator of religious estates in Khorasan, second only to Mecca in the scale of pilgrimage), Farokhi-Yazdi (renowned poet, journalist, deputy), Puladin and Jahansuz (leaders of the nationalistic officers driven to antagonism by Reza Shah's cronyism), and so on.

The king was not only insecure in his reign, notably his loose grip on the south, where Bakhtiyari tribes held sway, hence, the elimination of Sardar Asad as a preemptive measure, but he was also concerned about succession. His age was estimated as seventy by American diplomats, though ten years less in official figures. Meanwhile, the crown prince was still in his teens. This was a result of the royal bloodline that had begun with a string of daughters. Thus, it had become a pattern that those who were head and shoulder above the rest, in the king's entourage, would eventually be eliminated. And this by the admission of one of his most loyal ministers (Taghizadeh), who had been shrewd enough not to shine. This was not a brilliant strategy on the part of a king who was out to lay the foundation of a long-lasting dynasty rivaling the Achaemenid and Sassanid; for his ideal was pre-Islamic Persia, the age of the all-powerful Persian Empires (Greek and Roman periods), but nonetheless that was the *modus operandi*.[5]

In terms of this latest episode, the Erani Affair, although its contour would emerge within fifteen months of the arrests, its details would come to light decades later, in 1993, with the surprising publication of *Parvandeh* ("The Dossier"), buttressed by a further publication in 2003 (documentary addenda to a biography of Taghi Erani, title translatable as—*Erani, Beyond Marx*). These two works include memoranda from the police, the prison, the judiciary, the king's office, and various ministries; in a nutshell, declassified documents of the Pahlavi era (or deliberate leaks) belonging to this case. However, as the author of *Parvandeh* warns, there are many gaps in the story, including missing documents. This was particularly pertaining to an individual arrested with the group who had had a prior history, charged with "suspicion of espionage" for the Soviets in Persia's Air Force. For this offence, he had been disciplined (locked up for eight months, released, and expelled from the force, for lack of evidence, 1933). This person was confirmed years later by his associates to have been an NKVD agent (Soviet intelligence, pre-KGB). At any rate, his political practice could not be interpreted in any other way.[6]

The period following the fall of Reza Shah (September 1941) had been the perfect occasion to vandalize government archives and rewrite history as it were. This was the period when a Soviet-infiltrated political party by the name of *Tudeh*, meaning "the masses," reigned supreme in the country. They had a mass organization across the country, deputies in the *Majles*, occasional ministers, a military branch embedded in the armed forces estimated at five hundred, and political-military allies in provinces proximate to the Soviet Union. This coincided with when the Red Army occupied the entirety of north Persia (1941–1946). This party was influential until the *coup* of 1953, which reinstated Reza Shah's son on the throne (his reign during 1941–1953 had been wobbly), with a military branch that was dismantled only in 1958.

Offsetting this sabotage of the documentary record, a number of the actors and victims of these events published their memoirs and reminiscences in the ensuing years, notably after the 1979 revolution, the fall of the Shah (the second and last monarch of Pahlavi dynasty), which clarified part of the context and some of its details. Also, in 2008, a historian specializing in the Iranian left published a limited set of documents he had obtained in the wake of the Gorbachev era, in mid-1990s, from the former Soviet archives, connected to this case. Thus, the contour of this case has a much clearer definition today, although important gaps still remain and shall remain until further documents surface from private collections or archives connected with Iran.[7]

The group of detainees came to be called the "fifty-three." The chain presumably starting with the smuggler and the three arrested—Taghi Erani, the Berlin-educated medical doctor, and the man with the female alias running the "motel," followed by another forty arrests. This evidently adds up to forty four, not fifty three, hence a deficit of nine.

As it turns out, the smuggler had been arrested three months before Taghi Erani, and during the intervening period from his arrest (February 12) to the arrest of Taghi Erani (May 8), another nine arrests were made (February 14 to May 5). Of these, eight were so-called small fish, but one was a big catch. He was the NKVD agent alluded to above, a pilot by profession, trained in Russia, in his mid-thirties, having spent some twenty years of his life in Russia, codename *Tambourak* ("little lute," he had several in connection with this group), and April 5, or few days thereafter, his arrest, which was a full month before Taghi Erani's arrest. This person's confession constituted a full-fledged report, containing the names of over fifty individuals along with the scope of their alleged clandestine activities and mutual relationships. The earliest memo found in his records is dated April 27, eleven days before the arrest of Taghi Erani, but in this memo, he indicates that "as I have explained elsewhere...." and the interrogator asserting "as we know from prior depositions....," hence, the truncated nature of his file. He had made a mailing on April 5 to the smuggler, who was in police custody since February, which was intercepted by the police; shortly thereafter he was arraigned.

Of the fifty or so individuals named in his report, some had already been arrested in connection with other groups, and some were simply out of reach, living abroad. But twenty-six of them ended up on the roster of the "fifty-three." The rest were names provided by three undercover agents, and extracted from those rounded up, by various coercion and scare tactics.[8]

This process brought in a number of individuals who were bystanders, estimated at nineteen, including a tailor in Tehran who had seven children, and had allowed one of his clients to use his address to receive mail; and the latter had used it for so-called party work, the tailor acting as an unknown conveyer (*peyk*). A similar thing happened to a cobbler in Gorgan, near Port Shah in the southeast corner of the Caspian Sea, with ferry service to Soviet Union. There were also an estimated nine who were not an iota more than bystanders. They had met once or twice with some of those implicated, had brief chats about ideology and politics, and found no interest to further pursue these political leanings. One was of *Bahai* faith, and this precluded him from associations of a political nature. Conversely, there were those who had had affiliations, but their identity was never divulged to the police. There was also one person with a passing involvement, who had been named, but managed to avoid arrest. He used personal connections and paid a hefty fee (three hundred *Tumans*, equivalent to three hundred dollars at that time, sixty sterling). In short, of those arrested, half are believed to have engaged in some level of political activity in the context of this group, half of them rather seriously.[9]

Police rhetoric at the time was that the ex-pilot, the NKVD agent, was arrested two days *after* Taghi Erani. This is what all the detainees believed, including Taghi Erani. Given that a good number of them knew Taghi Erani, and vice versa, in their eyes he was the most likely culprit. And in the eyes of Taghi Erani, the culprit was the smuggler who had showed up at his house with the police, starting off the whole chain; and this only because his friend, the doctor, had been so reckless as to show up at his house with this dubious character.

This distorted timeline was what the police wished to instill in the mindset of those arrested, to place Taghi Erani at the origin of police discovery, thus indicating that he had betrayed his followers. This would deal a psychological blow to the grassroots opposition to the regime; the NKVD agent acted as an accomplice in the scheme. This perception persisted for fifteen months, until it was lifted due to police error. The police handed over the interrogation reports to the prison office that read them off to the detainees as a group (August 2, 1938), in preparation for court proceedings. This was to avoid giving them access to the files, and to make it swift, as the trial was going to be imminent (though it came in three months, November 2–15, 1938, with forty seven of the defendants, the remaining six tried later). However, even after this collective reading session, the detainees did not discover that the principal informer had been arraigned before nearly all the others, as the fabricated timeline was reiterated (only four had been arrested before him, the smuggler and the smuggler's brother being the first two).

The concept of trial for political prisoners was a novelty in Persia, invented with this group. Those rounded up on political ground never saw a trial, and it was uncertain how long they would stay locked up, some would pass away in prison without ever seeing a judge. Tehran was reputed to have one thousand such inmates, Tabriz six hundred, and other major cities, each several hundred.[10]

For this case, it had been decided that there would be a trial (the king, his Chief of Police and Justice Minister, the prime movers, deliberations lasting seven months, with official indictment issued in December 1937). Further, it would be a "big affair," hinting that the importance of this case was far greater than its intrinsic characteristics justified. Certainly, there was an element of foreign consumption to it, as well as domestic. Indeed, as records reveal, it was a central piece in geopolitical posturing of Reza Shah during the late interwar years, in which Persia's relations with the Third Reich figured prominently, as well as her relations with Bolshevik Russia and Great Britain.

In some sense, it was Persia's Dreyfus Affair (France), but with peculiar characteristics that made it the antithesis of this affair. The Dreyfus case had centered on espionage as the principal charge, whereas this affair was marked with deliberate downplaying of such charges and its total absence in the final sentencing (despite clear evidence proving act of espionage on the part of the ex-pilot and his associates). This was to avoid retaliation by the Soviets, a behavior for which there was ample precedence. Instead, the case would zoom on an intellectual who was an insider yet orthogonal to the existing order, someone by the name of Taghi Erani, and it would be a show-trial *par excellence*. There was also a good deal of what Orwell describes in *Homage to Catalonia* (1938) in the context of the Spanish civil war, how Moscow was more adamant on destroying the independent left than crushing the so-called stooges of world imperialism. With the latter, a deal could always be struck in exchange for *entente*, but with the former no such thing was possible. And in this case, Moscow's man did what he could to eliminate this figurehead of the independent left (after extracting from him his network, achieved right before the NKVD agent fell into the police net), not only to serve his masters in the Kremlin, but also his captors in Tehran; collaboration with the police in exchange for leniency and potential barter with anti-Pahlavi activists held hostage in Stalin's Gulags. These were the Persian operatives of the Comintern that were politically spent, good for barter or elimination, keeping them was only liability.

The petty informer who had taken the police to the residences of Taghi Erani and the doctor had been picked up in Ahvaz three months prior to the mass arrests, at the finale of his theatrical performance for working-class spectators. He was cuffed up behind the stage after the conclusion of his play called *Khiyanat va Vafa* (*Betrayal and Loyalty*). After some resistance, he had divulged what he knew, but little of what he knew was actionable (only the name and address of the doctor in Tehran). Further, as he was fearful of the Ahvaz dungeon, for it was reputed to be a *faramoosh-khaneh* ("house of the forgotten"), he insisted on being taken to Tehran, so he could show them the other locations by sight and identify the individuals. This led to a lengthy correspondence between Ahvaz and Tehran, until this course of action was approved. He was escorted to Tehran on May 6, 1937, two days before the arrests of Taghi Erani and the doctor.

Importantly, during the course of his haggling with the police, he had offered to make arrangements for a meeting with the ex-pilot (by sending a letter from the province to the doctor's address in Tehran, asking for an appointment, requesting a meeting with *Tambourak*). But the police did not take him up on his offer. By then, *Tambourak* was in their custody and cooperating in earnest. The mystery remained as to why collaboration with the intelligence service of a regime reputed as a British pawn in Soviet circles had not cost him his career in Soviet intelligence, or his life; rather, it had catapulted him to higher echelons. In his subsequent career, Stalin intervened personally to have him rescued from Persia (according to memoires of Ilya Dzhirkvelov, NKVD agent involved in the operation, 1948 episode). What subtle game had he played to secure such an outcome? And why had he gone beyond the call of duty in the scale of his revelations to Reza Shah's police?[11]

The answer to this is not so straightforward, but the simple part is that he did serve his masters by safeguarding the identity of Soviet infiltrators in Persia's army; by divulging the identity of dispensable pieces as a chess-like gambit, indeed pieces that had to be sacrificed, he had preserved the most valuable pieces. But there was more to this maneuver, as we shall discover.

The question also remained that if this group acted as a professional underground organization, how come no one took precautionary measures when their principal contact dropped out of view? Taghi Erani stated that the last time he met this character was after *Norooz* (Persian New Year, start of spring, late March), more than a month before his arrest. And further, how had this character fallen into the net cast by the police, especially since no one in this group seemed to know his name, nor where he lived, and he used multiple aliases. They could not even know if they were interfacing with the same individual; and even his mail to the smuggler was certainly from an ambiguous address.

There is also the inkling that, notwithstanding this individual's collaboration with the police—was not Taghi Erani suspected of covert activism from the outset of his return to Persia, given the precarious circumstances that had led to his departure from Berlin? Was he not marked from day one? And if so, why did this ending not come about earlier, especially since some of his writings after his return were rather thinly veiled in terms of their political implications.

Or was it a case of political rivals who were out to eliminate him, characters such as the Minister of Justice (Ahmad Matin-Daftari), who took an active role in the organization of the trial and threatening the lawyers who showed a semblance of courage in defending their clients (he became Prime Minister shortly after completion of the show trial). And was not the invitation to the *Golistan* palace another example of the Shah's obsession with observing his victims before passing on to the act (and if the royal invitations were declined that in itself would be proof of disloyalty). These, a string of questions and conjectures, signify a complex and convoluted case.

When Reza Shah eliminated Sardar Asad, his War Minister, he asked him to hand out the prizes for the equestrian competition they had just watched together, as he had to leave for some urgent matter. The War Minister, jubilant of this honor, handed out the prizes, was arrested and executed shortly thereafter. Reza Shah was fond of such games and played a similar trick with Teymurtash, when he asked him to return to the capital from his Caspian resort to resume his duties as Court Minister (effectively Chancellor), as the king's fit of anger was over.[12]

What motivates delving into the Erani Affair, besides the fact that the wealth of documentary evidence now allows a relatively comprehensive rendering, and the fact that it elucidates the *modus operandi* of Pahlavi in its founding and formative phase, which eventually led to its downfall and the birth of a political order which the west grapples with to the present, is that it sheds light on an episode that is highly revelatory of the mechanics of geopolitical posturing in the late interwar period.

That is, when appeasement with the Third Reich was pursued not only by England (Chamberlain, Munich Agreement) but also by Russia (Molotov-Ribbentrop), leading to preponderance of German presence in Persia. In effect, without transit routes controlled by Great Britain and Russia, the Persian Gulf and Volga-Caspian, trade with Persia was virtually impossible (Ottoman trade being minuscule). Therefore, Germany became Persia's largest trading partner, monopolizing nearly half of its trade, with windfalls in the short term for Reza Shah and his entourage, and prospects of dividends in the long term, security and stability, the latter the king's presumption and grave miscalculation.

Reza Shah's greatest fear had always been Russia. He knew them firsthand, having been a Cossack for over three decades in his years of obscurity (a force under Russian command in Qajar Persia). And as he was uncertain of British resolve in case of a Russian onslaught (across a common border stretching 2000-km), appeasement with the Third Reich, suggesting a general unwillingness on the part of the British to engage in conflict, he decided to hedge his chances with Hitler. And this he did, not only by subscribing to Schacht's vision of collective autarky, economic barter at large scale (raw materials for industrial equipment and infrastructure), but also by creating an ensemble of symbolism to highlight the affinity of his regime with the Third Reich.

This included the country's name change from Persia to Iran (land of Aryans, 1935) and swastika masonry in the capital's central train station (designed by a German architectural firm, Philipp Holtzmann), Boy Scouts mimicking Nazi youth camps (with a visit from Baldur von Schirach), Aryan propaganda (*Iran'e Bastan*, Ancient Iran, journal funded by Goebbels), and anti-Arabism in historiography and linguistics, as Persia's version of anti-Semitism. The persecution of communists, socialists, and anyone left of the political spectrum (trade-union activists) was another element of this strategy of show of unity with the Third Reich. This was driven by *Realpolitik* not ideological conviction.

What Reza Shah missed is that the policy of appeasement on the part of the British came to its end. The start of the war led to a psychological revolution and political upheaval in Whitehall (the downfall of "guilty men," and the advent of Churchill premiership). The subsequent course of events, notably Hitler's attack on Russia, led to the revival of the Lloyd George doctrine minted during the previous war—that Germany must not be allowed to devour Russia, as German know-how and Russian natural resources would be a lethal combination, the dawn of a German Europe, Europe a German Commonwealth.[13]

Hence, Russia had to be saved at all costs, and what better way to do this than through Persia. Persia had easily accessible ports (the Persian Gulf where the British fleet were already stationed, Bushehr and Port Shahpur, in contrast to half-frozen Archangel in the north, and far-flung Vladivostok in the east), and a Russia-bound railroad (TIRR, connecting the Persian Gulf to the Caspian Sea, a ferry ride to Soviet territory) ironically built with the intent to one day contain the Bolsheviks. Persia also had a king who had continued to flirt with the Third Reich even after appeasement was over. This was all of sixteen months, from Churchill's premiership to when the Allies invaded Persia, May 1940 to August 1941; the tipping point was when Churchill prevailed over Halifax in the war cabinet, May 28, 1940.

Disentangling the Erani Affair requires disentangling his life. His was a brief life (he was thirty-seven when he perished in prison) but traversed four political regimes—Qajar, Weimar, the Third Reich, and Pahlavi. His life became intertwined with these, the details of which have been emerging over the years, the archival disclosures and eyewitness accounts indicated earlier, the latest being family archives which have recently come to light (through the children of his first and third sisters, second sister had no children), which makes this project a technically viable undertaking.

Thus, what we have before us is a thematic biography at the cross-roads of four political orders that withered or came into being during the tumultuous interwar years, with its rich tapestry of cultural and intellectual life, a period with a certain resonance to contemporary history and its interwar-like rumblings. And as an added dimension, Taghi Erani articulated a certain political philosophy of modernism for trailing nation-states that provides a useful prism to view the subsequent evolution of historical events since his demise, epitomized by the fall of Pahlavi.

NOTES

1. Interogators were Javanshir (lead), Nosratollah Esfandiari, Ghodratollah Karami, Ali Mahboobin, Abbas Khan Fotoohi, Khosravani, and Parto Alavi (translator).
2. Early Soviet defectors, Bajanov (1928) and Agabekov (1930), former defected via Persia, latter with strong intelligence links to Persia divulged to the British. See their personal accounts, listed in bibliography, as well as account by Brook-Shepherd, *The Storm Petrels*.
3. See Pasyan and Motazed, *Zendegi'e Reza Shah*, pp. 379–380, for remarks attributed to Firouz.
4. For remarks of American diplomat Charles Calmer Hart about Teymurtash's brilliance and genius, see Majd, *Great Britain and Reza Shah*, pp. 174, 185.
5. See Afshar, *Zendegi'e Toufani*, p. 232, for Taghizadeh remarks about how Reza Shah eliminated capable ministers to protect succession in his bloodline.
6. See Momeni, *Parvandeh* (1993) and Boroujerdi, *Erani Faratar az Marx* (2003). The NKVD agent was Abdol-Samad Kam-Bakhsh. See Dehbashi, *Khaterat'e Siyasi'e Iraj Eskandari*, pp. 222, 267–268, and Boroujerdi, pp. 276, 514, 574, for remarks by Iraj Eskandari concerning Kam-Bakhsh's connection with Soviet intelligence. For similar assertion by Jahan-Shahlu-Afshar, see Boroujerdi, p. 495. For assertion by Khame'i, see *Panjah Nafar va Se Nafar*, pp. 38, 49.
7. For findings in Soviet archives concerning the Erani affair, see Chaqueri, *Taghi Erani dar Ayeneye Tarikh*, pp. 87–126, 158–179, 254–259, and 263. These pages contain facsimiles of original archival documents.
8. See Boroujerdi, *Erani Faratar az Marx*, p. 227, regarding Kam-Bakhsh's confession to Avanesian that twenty six of fifty three were named by him. Also, pp. 250–251, Amir-Khosravi asserts having seen the Kam-Bakhsh

report, and that twenty eight of the fifty three were named by Kam-Bakhsh. Khame'i, *Panjah Nafar va Se Nafar*, p. 100, says Kam-Bakhsh named twenty seven of the fifty three. Eskandari in Dehbashi, *Khaterat*, pp. 267–268, says Kam-Bakhsh had named twenty five to thirty of the fifty three and made it appear as if Erani had done so.

9. Dr. Ahmad Emami, college friend of Taghi Erani, paid off the police to avoid arrest.

10. Political prisoner population numbers given in the journal *Peykar* or Combat (Berlin and Vienna, 1931–1932); see reprints in Najmi Alavi, *Sargozasht'e Morteza Alavi*.

11. See Dzhirkvelov, *Secret Servant*, pp. 66–74, for his account of Kam-Bakhsh in Soviet intelligence operations in Persia in the 1940s and Stalin's direct intervention on his behalf.

12. For account of elimination of Sardar Asad and anecdote about equestrian race, see Pasyan and Motazed, *Zendegi'e Reza Shah*, pp. 575–576.

13. See Stone, *World War One*, p. 28, for description of Lloyd George doctrine; that Germany must not be allowed to annex Russia. "Guilty Men," denounced fifteen British politicians that had resorted to appeasement vis-à-vis the Third Reich; chief among them Chamberlain, MacDonald, Baldwin. Book came out anonymously and became a bestseller (June 1940).

CHAPTER 2

Tabriz (1902–1912)

Taghi Erani was born in 1902 in Tabriz, on March 21, on the eve of *Norooz*, the Persian New Year, the start of spring, the vernal equinox.[1] On the Persian calendar, this is a leap day (30 *Esfand*, a month of twenty-nine days) and occurs once every four years. He was the first child of a couple who happened to be cousins, his parents' fathers were brothers.

The father of these two brothers, the great grandfather of Taghi Erani, is the most distant progenitor of the Erani family of known identity, born around 1800 in Qaraje-Dagh, a mountainous region 100-km east of Tabriz. His name was Hajji-Mulla Ahmad Mojtahed Qaraje-Daghi. His ancestry is uncertain, but several pieces of information point to the Caucasus region.

The name Erani was adopted by his descendants, his grandson, the father of Taghi Erani. Eran is in the southeast of Caucasus, corresponding to the present-day Republic of Azerbaijan. Adam Olearius passed through this region in the 1630s and provides a description of Eran as the region enclosed between the rivers Kura and Aras. There is also panegyric poetry of Qatran (c.1050 AD), praising the rulers of Eran.[2]

Second is the known history of this region. The Russian objective since the late eighteenth century (reign of Catherine the Great) was to gain access to the Black Sea, which led to the loss of the Caucasus by indigenous tribes (Circassians), losses by the Ottomans (Crimea on the northern coast of the Black Sea, and Batumi province, the southeast corner of the Black Sea); and finally, losses by the Persians (Georgia, Armenia, Azerbaijan, in the south Caucasus).

© The Author(s) 2019
Y. Jalali, *Taghi Erani, a Polymath in Interwar Berlin*,
https://doi.org/10.1007/978-3-319-97837-6_2

The Caucasus is a landmass the size of Sweden (twice the British Isles), split into northern and southern parts, separated by a mountain range and running along the axis that connects Sochi to Baku. It touches the Caspian Sea to the east and the Black Sea to the west. The latter is far more important, as it is a direct link to Europe and is connected to the open seas via the Mediterranean. The Caspian, on the other hand, is landlocked in Central Asia, and only called a sea by virtue of its size.

The Persians lost their possessions in the south Caucasus to the Russians, who undertook a policy of demographic fortification with "Christian stock," extending what had been achieved by the Romans and Crusaders in Georgia and Armenia. This was achieved by bringing in fresh settlements of Volga Germans and other citizens of the vast and growing Russian Empire. This led to the renaming of certain cities, including Ganja (became Elizavetpol, in honor of the Tsar's wife). Ganja had been the native city of the master of Persian romantic poetry, Nezami Ganjavi (d. 1209 AD):[3]

> *Now from the goblet, ruby wine they sip*
> *Now interchange their kisses, lip to lip*
> *Now hidden mysteries of love unfold*
> *Now in close embrace, each other hold!*

The loss of Ganja to the Persians is somewhat like the English losing Stratford-upon-Avon, Shakespeare's birthplace. These losses triggered the migration of those of Persian heritage from the southern Caucasus to present-day Iranian Azerbaijan. The border is demarcated by the Aras River. The bulk of this migration took place during 1790s–1830s. The parents or the guardians of the great-grandfather of Taghi Erani, therefore, may have been among the first wave of such a migration.

The great-grandfather had a name that was not only indicative of his township (Qaraje-Dagh), but also of his faith and profession; *Hajji* means that he had made the *Hajj* pilgrimage to Mecca. This involved travel in a *caravan*, passing through many towns and cities in Mesopotamia and the Arabian Peninsula. Such a journey would have taken several months. In the context of those days, he would be considered a worldly traveler. The name *Mojtahed* indicates he was not a mere *Mulla* (someone with rudimentary knowledge of theology), but a learned scholar of Islamic jurisprudence or *feq'h*. He may have been a *Mulla* in youth and a *Mojtahed* later in life, compounding the two as his identity in the community.

The family genealogy identifies him as the tutor of the youngest children of Prince Abbas-Mirza (d. 1833), who was poised to become the third monarch of Qajar.[4] However, he predeceased his father; thus, his eldest son acceded to the throne, Mohammad Shah (r. 1834–1848). The great-grandfather of Taghi Erani, therefore, was the tutor of the youngest siblings of Mohammad Shah. This tutoring position meant that at some point, possibly since 1820s, the great-grandfather had moved from Qaraje-Dagh to Tabriz, the provincial capital of Azerbaijan, the seat of the crown prince.

Hajji-Mulla Ahmad had seven children, two daughters and five sons. Two of these sons, among his youngest, born in the 1840s or 1850s, lead us to the Erani family. One of these sons, Mohammad-Ali, was a *Mojtahed* like his father, a theological scholar. And his daughter, Fatemeh (b. 1881), was the mother of Taghi Erani; though in appearance she seemed free of strict codes of Islamic garb (several photographs have reached us, as well as testimony of one of her grandchildren), she maintained definite religious beliefs.[5]

The other son was named Taghi Amin, but his profession is unknown. Taghi means one who refrains from indulgence, best rendered as pious, and his surname, Amin, means trustworthy. Aliases were common in those days, and most were pegged to social standing, which tended to flatter those possessing greater means with greater moral virtues. At any rate, this son was the father of Taghi Erani's father. Taghi Erani's father was born in 1882, and had the alias Amin Hazrat (Trustworthy Sire), and was rather well-to-do. There is a photograph in the family archives from the 1920s, showing him chauffeured in an automobile.

Amin Hazrat was associated with the provincial administration of Azerbaijan, the tobacco bureau, which for a short while had been the tobacco state monopoly or *régie* (1891–1892). After this short-lived *régie*, the bureau managed the crown lands where this crop was cultivated. Tobacco was a major cash crop, grown on large estates which included crown lands, endowed lands of religious foundations (*o'ghaf*), and lands of absentee landlords, all of whom needed men with knowledge of local customs to carry out collection of the ground rent. This was typically four-fifths of the total produce, collected in kind from indentured laborers. These were called *ra'i'yat*, and were no better off than serfs, and had duties beyond tilling the land to which they would be tasked by their lord's agents (*mobasher*). They comprised the great majority of the rural population and lived wretched lives.

The population of Persia at the close of the nineteenth century was about twelve million, of which a third was urban, a third nomadic, and a third rural; amongst the rural population, the serfs were the great majority, the rest were petty landowners, artisans, day laborers, and peddlers.

The cash part of cash crop came in when the produce was monetized on the market, including export markets, through the chain of local dealers and merchant trading houses. Besides tobacco, cotton and opium poppies were the two other cash crops in those days. When the civil war in America caused a decline in the export of cotton to England, this was partly compensated by cotton from Persia.[6]

It is reputed in the oral tradition of Persian folk culture that cousins have their union prefigured in the heavens. This is not something which one openly admits in the prevailing culture of our time, which shuns consanguine marital relationships, but it is nevertheless observed even to this day in Iran. And certainly, during that time period, it was rather a common occurrence. In a way, it was the lesser of two evils, for if giving away a daughter bore a high chance of her becoming a domestic maid to the hordes of in-laws, the risk of such mishap could be mitigated by dealing in circles about which one had greater certainty and some leverage, such as amongst relatives.

Thus, Fatemeh was given away to her cousin, Amin Hazrat, a union that took place around the year 1900 in Tabriz, the provincial capital of Azerbaijan, seat of the Qajar crown prince and hence, the second most important city after the capital Tehran. From this wedlock came five children—two boys and three girls. Taghi was the first (b. 1902), then came Hashem (b. 1904) who perished at the age of two. Infant mortality was common in those days, and parents often would not record the birth until the newborn had passed the winter and proven to be of sturdy stock. (There was no national registration in Persia before 1920s, but births were recorded by traditional means, in a local *mahzar* or notary, or by inscription in a holy book or *divan* of poetry, witnessed by a local notable.)

After the boys, came the girls. The first was rather romantically called Irandokht (b. 1909), meaning daughter of Persia, later a feminist activist with a literary history.[7] This is not such an uncommon name for a female, but coming from an ethnic minority, it suggests that the family did not have a strong provincial inclination but rather a national one. The people of Azerbaijan have their own language (Azeri), which is Turkish with admixtures of Persian, and this was the home tongue of the Erani family. This situation would be similar to someone from Wales, Brittany, or Catalonia identifying themselves as British, French, or Spanish, although they speak their ethnic tongue at home.

After Irandokht (b. 1909) came Kokab (b. 1911), which is the name of a species of dahlia common in Persia, and then Shokat (b. 1920), which means magnanimous. Shokat is the only one of the siblings who was not born in Tabriz; she was born in Tehran.

The earliest photograph of Taghi Erani in the family archives is when he was six. He is wearing a cap and bowtie, smartly dressed, possibly for *Norooz*. The photograph is most likely from March 1908. Shortly after this, civil war broke out in Persia, and Tabriz became its epicenter. This embroiled the city for over a year, a dark episode from June 1908 to July 1909.

This conflict was not about the overthrow of monarchic rule in Persia, but whether it would remain a constitutional monarchy (as conceded by the king in 1906, now deceased) or if it would revert to absolutism (as his son wished to bring about). Absolute monarchists called themselves royalists and were backed by Russian troops (as the king was their marionette), while the constitutional cause was represented by a broad movement of those who called themselves nationalists. This included a wide spectrum from moderate reformists to radical revolutionaries, reflecting the diversity of grassroots Persia.

The royalist–nationalist conflict reached a stalemate; therefore, political tensions did not subside, and in December 1911, Tabriz witnessed another dark episode, inflicted by Russian troops against constitutionalists. This triggered the departure of the Erani family from Tabriz to Tehran, which was considered to be a safe haven, relatively speaking. The move happened in 1912, spring or summer, when Taghi had turned ten and had come to have two sisters, Irandokht and Kokab.

Another factor contributing to the move to the capital may have been the estrangement between Taghi's parents, and the material possibility of being able to take up residence in Tehran, due to a family inheritance on the mother's side; but this is conjecture. What is certain is that Taghi began primary schooling in Tabriz (*Fiyouzat* or Endowed, c. 1908). His first sister was born in Tabriz (1909), as well as his second sister (1911). Taghi completed primary schooling in Tehran (1914, *Scharaf Mosaffari*, Honor of Mozaffar), the school named after the Qajar monarch who decreed

Persia a constitutional monarchy. Taghi attended *Dar-ol-Fonun* in Tehran, the Polytechnic, for his secondary school (1914–1920), and College of Medicine (1920–1922), at the same premises as *Dar-ol-Fonun*, but with a parallel administration. Thereafter, he left for Berlin (August 1922).

The birth of Taghi's third sister, Shokat (1920), indicates that the relationship between his parents had not been severed, at least not irrevocably. The long separation between the births of the second and the third sisters (1911 and 1920) may be explained not only by the move of the mother and the children to the capital, but also by the outbreak of the Great War (1914–1918), which reached Persia soon after it erupted in Europe (June in Europe, October in Persia). This emblazed Azerbaijan and reached within 60-km of the capital Tehran (town of Karaj), rendering domestic movement during this period virtually impossible.[8] It was only after the end of these hostilities, we may surmise, that the family could reunite.

Western Persia became the battleground between an alliance of the Russians and the British on the one side, and Ottomans and German-backed guerillas on the other; the latter group pressing to capture the oilfields of Baku in Russian-controlled Caucasus, and oilfields of Khuzestan in British-controlled Persia. This was despite Persia's declared neutrality in the war.

Before we get to Taghi's life in Tehran, which became a springboard for his journey to Berlin, it may be useful to explain the events in Tabriz that triggered the dislocation of the Erani family to Tehran.

TABRIZ UPHEAVALS (1908–1911)

In the *Sydney Morning Herald* of December 22, 1908, we come across the following news clip about Tabriz:

BATTLE AT TABRIZ. BOMBARDMENT AND SORTIE. NATIONALISTS ROUT ROYALISTS, WOMEN AS SOLDIERS. *Ain-ed-Dowlah, leader of the royalist forces before Tabriz, shelled the city, killing and wounding one hundred nationalists. The horsemen of the nationalists, under Sattar Khan, then made a sortie under cover of artillery, and routing the Shah's troops pursued them for thirteen miles. Twelve guns, and quantities of provision and ammunition were captured by the nationalists. The casualties on both sides were two thousand. The nationalist dead included sixty-seven women in male attire.*[9]

Above was dispatched by an English correspondent, Arthur Moore, who had been sent to Tabriz by the British press (co-funded by *Daily News*, *Daily Chronicle*, *Manchester Guardian*, and the Balkan Committee of the House of Commons) to report on the outbreak of civil war in Persia and break the Russian news monopoly on this important development. *Sydney Morning Herald* relayed this from the British press.

The year 1908, when these events broke out in Tabriz, triggered by events in the capital Tehran, is the year when Persia ceased to exist as a sovereign nation. This sovereignty was not restored until the revolution of 1979. That is, Persia went through successive periods of foreign domination—Anglo-Russian (1908–1917), British (1918–1941), Allied (1941–1953), including Allied occupation during 1941–1946 (Red Army included), and American (1953–1979); the last came to its end with the fall of the Shah.

Prior to 1908, for over a century, Persia had been retreating in the face of Russian expansionism and British infiltration, one encroaching from the north, the other from the south, with pacts between them to divvy up this sick man of western Asia; a mere shadow of its past, the Persian Empires of Greek and Roman times, the Safavid of pre-modern times.

The Russians were continuing with their drive to access "warm waters," the Black Sea secured, the Persian Gulf the next target; if not, then the northern provinces of Persia. The British, on their part, sought an uninterrupted chain of colonies from India to Egypt, their two great possessions in the Near East. This meant mastery of the Persian Gulf, hence mastery of south Persia; as well as mastery of Afghanistan, as a buffer against Russian assault on British India. This meant extracting from Persia what remained of her Afghan possessions, notably Herat.

During this colonial expansion, the Persian monarchs managed to maintain their independent line of action, although often with poor results. They did this initially by armed opposition and eventually by the bizarre balance-of-power strategy of giving away concessions to the two rivals in equal measure (for which the king and courtiers received hefty commissions). Meanwhile, they sought triangulation by trying to engage with a third great power, initially France, then Germany, to try to loosen the bipolar straitjacket.

The master of this juggling act was the fourth king of Qajar (as the first three were martial kings), the longest ruling Qajar monarch, Nasser-ol-Din Shah (r. 1848–1896), with his son Mozaffar (r. 1896–1907) following in his father's footsteps. Mozaffar is the one who gave in to pressure from Persia's elite—the powerful landlords, merchants, ecclesiastical authorities, tribal chieftains, and barons of bureaucracy—who used the revolutionary energy of the masses that had culminated in the assassination of his father to eke out concessions from the court in the form of a legislative assembly or *Majles* enshrined in a constitution. Thus, Persia's nobility and gentry, and partly its commoners, found a say in matters of the state. This happened in August 1906, the culmination of the constitutional revolution of Persia. But shortly thereafter, the constitution-granting king passed away (January 1907).

With his passing away, also passed away Persia's existence as a sovereign nation, a sea change that became manifest in 1908, a year after Mozaffar's son acceded to the throne, a marionette in the hands of the Russians. This king was resolved to restore the rule of absolute monarchy in Persia and had the luxury of having a Russian henchman at his disposal, General Lyakhov. Lyakhov was the head of the Persian Cossack Brigade (PCB), a force which was nominally accountable to the Persian monarch, but in reality, was accountable to Russia, with Russian officers and a Russian general at the helm, reporting to the Russian minister in Tehran and chain of command in St. Petersburg. Only the foot soldiers and part of cavalry consisted of Persians, all subalterns.

This upstart king (his name was Mohammad-Ali Shah, with the shortest reign of any Qajar monarch, thirty months, as opposed to his father's ten years, and grandfather's nearly fifty years) also had an advisor, who had been his tutor of Russian in his youth; Schapschal, a manipulator in the vein of Rasputin—an "evil genius" in the words of Professor Edward Browne.[10] And to complete the cast of characters, there was Baron Hartwig, the Russian minister in Tehran, and renowned Slav extremist.

Consequently, when the absolutist king assumed the throne and sensed the resolve of the nationalists to preserve the constitutional order (the radical wing of nationalists, run by a secret society, organized the assassination of his chancellor and made an attempt on the king's life), he decided to act forcefully. Therefore, the *Majles*, Persia's parliament, the symbol of her fledgling democracy not yet two years of age, was bombarded at the command of General Lyakhov. Given that the *Majles* deputies were largely advocates of moderate reform, the assault meant that there was no scope even for mild reform; despotism was all Persia deserved.

What saved the situation, as far as the nationalists were concerned, is that this king, prior to becoming a king, had been the crown prince and therefore the ruler of Azerbaijan province and its capital Tabriz. This meant that the people of Tabriz knew what sort of character they were dealing with. And Tabriz, being at the crossroads of Europe—connected to the Mediterranean (via Aleppo), to the Black Sea (via ports of Batumi and Trebizond), and to railways and land routes of the Russian and Ottoman Empires—had plenty of ideas about how the greater world and particularly Europe were evolving, and how this stood in stark contrast to the despotic mode of the government at home.

(The connection of Tabriz to Europe is ancient. In Gibbon's history of the Roman Empire, we read the following itinerary to Taurus, ancient name of Tabriz—"From Constantinople to Trebizond, with a fair wind, four or five days; from thence to Erzerom, five; to Erivan, twelve; to Taurus, ten; in all, thirty-two. Such is the itinerary of Tavernier who was perfectly conversant with the roads of Asia."[11])

Accordingly, the citizens of Tabriz took up arms under the command of a commoner named Sattar, a horse trader and former militia, and they drove out the royalist forces from their city. Sattar became a national hero overnight, outcast turned revolutionary (a forerunner of Pancho Villa), and he was hailed in the foreign press as the Garibaldi of Persia.[12] The royalists retaliated by choking off the city and taking it to the brink of starvation. The fighting in Tabriz broke out in June 1908, a few days after the bombardment of the *Majles* in the capital. The expulsion of the royalists was completed by October, and the last clashes when the nationalists had freedom of movement outside the city took place in December, as reported in *Sydney Morning Herald*.

Then, by January 1909, the siege took hold and Tabriz became a big prison. While Tabriz was slowly starving, the other cities rose in indignation and two contingents of nationalist forces, one from the north and the other from the south, made a dash for the capital Tehran, to unseat the renegade king and bring down the entire royalist order.

Meanwhile, Tabriz made heroic attempts to break the siege, involving legions of volunteers, including women. A young American missionary by the name of Howard Conklin Baskerville, a Princeton graduate from Nebraska, a pupil of Woodrow Wilson, and a teacher in Tabriz, was one such volunteer. He led a contingent of siege-breakers in an attempted raid, but fell to a royalist sniper's shot (the American consul in Tabriz unable to dissuade him from joining the effort);[13] this was in April 1909.

A few days later, at the end of April, Russian troops arrived in Tabriz from the Caucasus, on pretext of protecting their diplomatic corps. What allowed them to maintain this façade was that it was an impossibility for them to reach Tehran. The nationalist forces heading toward the capital controlled the strategic junction of Qazvin. Qazvin was the ancient capital (c. 1550–1600), where roads from the north, south, and west led to Tehran to its east, and it was an impossibility for the Russians (being in Tabriz in the far west) to reach the capital and save the king. The king could only count on his local Cossack Brigade. Thus, they stood by watching, while keeping Tabriz hostage, but not harming the city as of yet, and so, the pretext held up. Meanwhile, they disarmed Tabriz, but allowed the delivery of food to the city. The siege was lifted, after four months, the famous winter of fodder (town-folks reduced to consuming food intended for livestock, *yonjeh*).

From the time the Russians entered Tabriz, end of April, it took two and a half months for the nationalists to converge on the capital Tehran, break into its moated citadel, and following four days of street fighting, to force the abdication of the king (who took refuge in the Russian embassy). All along Tabriz was under Russian occupation, and they followed the events through streaming telegrams. Given that Tabriz was held hostage by the Russians, the nationalists who conquered Tehran did not dissolve the Cossack Brigade, and General Lyakhov was maintained in his post of PCB commander. Thus a stalemate of sorts was achieved. This was in July 1909.

It is impossible to know what was experienced by the Erani family in the midst of this conflict in Tabriz. What is certain is that sounds of musketry, artillery, mourning, and the feeling of uncertainty and desperation pervaded the atmosphere, followed by sightings of Russian soldiers, curfew, and patrolling operations. The experience was surreal for young and old alike. The leader of the insurgency, Sattar, and his associates, took refuge in the Ottoman embassy when the Russians entered Tabriz. They did not emerge until the fall of the renegade king in the capital and written assurances against retaliation.

In the meantime, a king had been anointed in the capital who was not much older than Taghi, the young Ahmad Shah, the eldest son of the deposed king, only eleven years old (Taghi was seven). The absolutist king, following his abdication in 1909, went to exile in the port city of Odessa with a generous state pension, while the child monarch was crowned under the guardianship of a Qajar elderly. Needless to say, no one took the child king or the guardian seriously. Therefore, the Russians made an attempt to reinstate the old king, in 1911, and when that operation backfired, they occupied northern Persia and brought about the dissolution of the *Majles* a second time.

The *Majles*, after order had been restored when the absolutist king was dethroned in 1909, had brought in American experts to create a financial administration in Persia and Swedish experts to create a national gendarmerie for the maintenance of law and order and tax collection. Immediately after the dissolution of the second *Majles* in 1911, and the expulsion of the American financial experts (led by William Morgan-Shuster, who wrote his memoires shortly thereafter—*The Strangling of Persia*), the Russians savagely stamped out the protest movements triggered in its wake. In Tabriz, they hanged thirteen prominent nationalists in public view (December 1911), and in Mash'had they bombarded the shrine of Imam Reza (March 1912).[14] Now, north Persia was de facto Russian-controlled territory.

In the intervening period of relative calm from March 1912, when the Russians flexed their muscles for the last time in Persia, before the Great War, to October 1914, when they returned following the eruption of the war in Europe, the Erani family left their native Tabriz for the capital Tehran. This was caused not by an anticipation of the Great War and its contagion to Persia, which could not have been known at the time, but by the inherent insecurity of Azerbaijan and Tabriz, as evident from the upheavals of 1908–1911.

This is the time frame in which Taghi's mother left for Tehran, accompanied by her children—Taghi (ten), Irandokht (three), Kokab (one)—with Taghi's father staying behind, although he may have escorted the family in the journey. This was in 1912, two years before all hell broke loose in Europe and its dire effects spread to the Near East and Persia.

NOTES

1. Two dates of birth appear in the documents with Erani signature (March 21, 1902 and September 5, 1902). The first is given in his *Lebenslauf*, when he applied for the teaching position at the University of Berlin, on January 27, 1925. See *Bankverbindung*, p. 59. Again, in February 1928, in the draft of his biography for the doctoral dissertation, he repeats this date. See Chaqueri, *Taghi Erani dar Ayeneye Tarikh*, p. 207. The date September 5, 1902, however, is given in his official university dissertation. This change suggests the original date is correct, but he was obliged to change it to be in agreement with his official documents (passport, based on his birth certificate, prone to late dating).

2. For history of Eran or Aran, see articles by Bosworth, and also Reza; for account of Olearius visit to Persia see Werner; also see Mansouri's work on history of Azerbaijan, Vol. 1, p. 326; see Botting for panegyric poetry of Qatran.

3. See Browne, *A Literary History of Persia*, Vol. 2, p. 408, for Ganjavi verses cited in the text.

4. Geneology obtained from *Erani Family Archives*.

5. According to the eldest child of the youngest sister of Taghi Erani (Shokat), who remembers his grandmother in early 1940s (she passed away in 1946), she wore neither the *chador* (veil) nor other forms of *hejab*; personal communication, August 2016.

6. See Jamalzadeh, *Ganj'e Shayegan*, p. 17, on increased export of Persian cotton to England during the American civil war.

7. See literary works of Irandokht in the bibliography, three works in Persian, one in English.

8. See Miroshnikov, *Iran dar Jang'e Jahani'e Avval*, pp. 54–55, describes how the forces of General Zolotaryov advanced to Karaj, 60-km west of the capital, and how this led to the flight of Persia's political class from the capital to Qom and later Kermanshah, first occurrence, October 1915.

9. See *Sydney Morning Herald* in archival section.

10. See Browne, *The Persian Revolution of 1905–1909*, p. 198: "Shapshal Khan, the Russian Jew who was the Shah's tutor during his youth, and throughout his reign his evil genius."

11. Gibbon, *The History of the Decline and Fall of the Roman Empire*, Vol. 5, Ch. 46, Footnote 84.

12. See Browne, *The Persian Revolution of 1905–1909*, pp. 441–442 on Sattar as Garibaldi.

13. See remarks attributed to Howard Baskerville in Calafi et al.: "The only difference between me and these people is my place of birth, and this is not a big difference."
14. On December 30, 1911 Russian forces in Tabriz hanged Seghat-ol-Eslam and twelve other nationalists in Bagh'e-Shomal. On March 30, 1912, Russian forces in Mash'had shelled the Shrine of Imam Reza. Both were measures to suppress unrest in the wake of the dissolution of the *Majles*, which happened on December 24, 1911.

CHAPTER 3

Tehran (1912–1922)

At the turn of the century, the state of long-distance transport in Persia was rather abysmal. While in Europe and Eurasian powerhouses (Russian and Ottoman Empires), railways were rapidly growing, in Persia, besides a ten-kilometer leisure line, a tramway from the capital Tehran to the historic town of Rayy, there was nothing else but the horseback and the *tarantass*, a Russian-design carriage reputed for its stiffness. (Rayy was home to a well-known shrine and historical relics of Seljuq times, eleventh and twelfth century; it was a favorite promenade of Qajar kings.)

Further, these primitive modes of transport had no relaying stations other than what had been inherited from the Persia of antiquity: the *chapar*, where fatigued horses could be traded for fresh ones, and the *caravanserai*, where travelers could spend the night, people and transport animals.

Friedrich Rosen, who we shall get to know when Taghi Erani gets to Berlin, traversed Persia from the Persian Gulf to the Caspian Sea on horseback in 1887. In his memoires, he recounts that he much preferred the horseback to the *tarantass*, as it had no shock absorbers, nor cushioned seating, only a stack of hay as compensation.[1]

Rosen spent eight years in Persia in the 1890s, as German minister, and vividly describes the state of Persia's mode of transport at the close of the nineteenth century. Then came Persia's civil war and the Great War, precluding improvement in the condition of domestic transport until the 1920s.

© The Author(s) 2019
Y. Jalali, *Taghi Erani, a Polymath in Interwar Berlin*,
https://doi.org/10.1007/978-3-319-97837-6_3

Physical discomfort, however, was the least of worries of a people who were brought up in ruggedness. Physical danger in the form of banditry was much more of a concern. For this reason, long-distance travel was limited to the spring and summer months, when the climate was more accommodating, and the days longer, making it possible to do more distance in daylight. Night travel was to be avoided at all costs. Grouping several carriages to create a *caravan* (the horseback was for solo travelers, or groups with a light load), escorted with armed men, and routing the trip through itineraries dotted by chain of *caravanserai* and *imamzadeh* (shrines of minor saints) were recipes for a safe journey.

This is all we can know about the logistical aspects of the journey of the Erani family from Tabriz to Tehran in 1912. The distance was 650-km, with the city of Zanjan representing the midway point of the journey. From Zanjan to Tehran, the midway point was the city of Qazvin, which had been the second Safavid capital (c. 1550–1600) and was literally the high point of the journey (1800-m altitude, Tabriz at 1300-m). And midway from Qazvin to Tehran was the town of Karaj, with its two streaming rivers (Kordan and Karaj) flowing from the foothills of the Alborz mountain range in the north of Persia. Beyond Karaj, it was a two- or three-day journey to a vantage point where one could hold the citadel of Tehran in one's full view. The entire journey would take two to four weeks, depending on the circumstances.

Tehran was quite different from Tabriz in a positive way, although the two had comparable populations (estimated at 150,000), and it was Tabriz that had the more illustrious history, going back to antiquity. But in terms of recent history, it was Tehran that held the edge, since Nasser-ol-Din Shah (r. 1848–1896), the longest ruling Qajar monarch, had done much to modernize it and emulate aspects of European architecture. He made three trips to Europe—1873, 1878, and 1889.

But everything in Persia was a hodgepodge, so whatever sense of European utility and aesthetics was imitated, the backbone remained Persian and Perso-Islamic. For one thing, the medieval idea of city as citadel was still the chief idea. Tehran, which came to fore when the Safavid made Qazvin the capital (c. 1550), as a fallback position in the event of an Ottoman attack, it dwindled when the Safavid moved the capital to Isfahan in the center of Persia (c. 1600). However, Tehran bounced back when the founder of Qajar made it the capital of Persia (1789), and it underwent massive expansion under Nasser-ol-Din Shah (r. 1848–1896).

Tehran leaped from a walled (*hesar*) barracks three kilometers in perimeter, to a walled citadel, twenty kilometers in perimeter, with one hundred and fourteen watchtowers (the number of *Suras* in the Koran), a massive moat, a dozen gates (three on each edge of a square-like fortress) leading to retractable bridges over the moat. There was pedestrian and mounted traffic through the gates, and collection of levies from those bringing merchandise into the capital. It operated from dawn to dusk, after which it was shut as night curfew fell.[2]

Outside the citadel lived the moat-dwellers—odd nomads, herdsmen, gypsies, dervishes, delinquents, dealers, prostitutes, and outcasts of all stripes—as well as travelers who had arrived after the curfew and had to tent overnight to gain access to the citadel the following morning.

Inside the citadel, the walled barracks of Safavid period (*Arg*) had become the exclusive walled royal quarter of Qajar kings, with a palace (*Golistan*, rose garden), the Armory (*Ghoor-Khaneh*), the Academy or Polytechnic (*Dar-ol-Fonun*), and administrative buildings. Outside the royal quarter (public space) there was the Royal Mosque and the Grand Bazaar to the south, and the Royal Boulevard (*Bab'e Homayoon*) to the north, the latter inspired by European avenues with sidewalks, gas lighting, greenery, and stalls of merchant trading houses.

These sectors comprised the central strip of the citadel. Westside consisted of neighborhoods for military personnel and the bureaucracy. Eastside was less affluent with ethnic- and guild-designated neighborhoods, and traditional outlets of urban commerce and public space. This included bathhouses, teahouses, bardhouses, wrestling houses, water tunnels (steeply sloping stairways to reach underground points of potable water), domed ice stores, orchards, *caravanserai*, and cemeteries. There were many narrow alleys and four-way junctions (*charsoo*), and the main roads were cobblestoned, flanked with rivulets which ran downhill north to south.

Mansions and villas of nobility were outside the citadel in the northern suburbs. Beyond these affluent suburbs, one reached the foothills of the Alborz mountain range, with the snow-capped Mt. Damavand overlooking the city. It was very different than Tabriz which had a rustic feel, with a river flowing through it, old stone bridges, mud-thatched homes, and narrow alleys, in a valley with a view of two opposing but gentle mountains.

The Erani family settled in the westside of the citadel, in *Sheikh Hadi*, two kilometers from the royal quarter and the elite Academy to which Taghi would soon gain access. They hired an errand boy (*gomaushte*), a sign of the family's affluence, who was sixteen years old at the time. He was six years older than Taghi. He went on to work for the Erani family for over a quarter of a century, raising his own family with three children, and we see his trace as the life of Taghi Erani begins to eclipse. His name was Mash'hadi-Reza Damavandi.[3]

Having an adult male to run errands in a male-dominated society was certainly an important factor in surmounting the challenges of daily life in the capital. The fact that Tehran had a large population of ethnic Azeris must have also facilitated getting settled in the new environment. Ethnic solidarity is a strong cultural trait in Persia. And quite by chance, all this had happened before the flames of the Great War spread to Azerbaijan, the home province of the Erani family, where the father stayed put, carrying on in the tobacco business.

DAR-OL-FONUN, POLYTECHNIC OF PERSIA

Taghi's schooling had begun in Tabriz, by all indications four years of primary schooling at a school named *Fiyouzat* (Endowed), 1908–1912, though we do not know how the civil war in Tabriz affected the start of his schooling. In this period and possibly from preschool age, he was also put in Koranic classes, at the instigation of his mother, where he developed a sound knowledge of Arabic. This was to serve him well when he arrived in Berlin, as it became the basis for his collaboration with Friedrich Rosen.

In Tehran, he completed the last two years of primary education at a school named after the Qajar monarch, who decreed the establishment of constitutional monarchy in Persia, Mozaffar-ol-Din Shah. The school was called *Scharaf Mosaffari*, as transliterated by Taghi Erani in accordance with German phonetics, meaning Honor of Mozaffar, and was founded by an enlightened Qajar nobleman (Ehtesham-ol-Saltaneh) who pioneered modern primary schooling in Persia.

It should be noted that at that time there was neither universal education nor secular schools as such, only the traditional *Maktab*. And even in the modern schools established by the educated elite of Persia or missionaries, the teaching of traditional sciences was still the norm, such as Ptolemaic astronomy with its geocentric dogma. When Copernican astronomy was first taught in Tabriz, it was indeed big news; that was in 1906.[4]

This lack of proper academic institutions was an aberration that progressive actors of society wanted to remedy. It was embarrassing for a country that prided itself on having had world-leading institutes of higher learning in antiquity and the Middle Ages (*Gunde-Shapur, Nezamiye*), leading such fields as medicine, mathematics, astronomy, and chemistry, with a cast of scientists who were household names in Persia, and known to savants elsewhere (Khwarazmi, Rhazes, Biruni, Avicenna, Khayyam, Tusi) that it had descended into the abyss.

Abbas-Mirza (b. 1789), who as crown prince was poised to become the third Qajar monarch, but died a year before his father (in 1833), began this process of renaissance by sending students to Europe as early as 1813 to survey the state of knowledge and report back as to what ought to be done. His motivation was to modernize the military, that is, the provincial army of Azerbaijan (as the crown prince, he was the governor). He had the Russians breathing down his neck, and modernization of military was a matter of life and death.

Under his watch, the production of armaments and ammunitions, the development of artillery, and engineering logistics to support military campaigns made certain strides. But in comparison with what the Russians had achieved (since Peter the Great, d. 1725), it was too little and too late. For this reason, the southern Caucasus was lost to the Russians by 1828, after three decades of on-and-off warfare. The textile industry spurred by mass production of uniforms for the soldiers also made advances. In these initiatives, Abbas-Mirza was advised by the father and son team of Ghaem-Magham, forward thinking counselors to the crown prince.

This line of thinking developed further under Amir Kabir, who became the chancellor of Persia for a brief period of three years (1848–1851), when Nasser-ol-Din Shah, the grandson of Abbas-Mirza, succeeded to the throne. Amir Kabir had been a protégé of Ghaem-Magham Senior and Junior and developed the idea of establishing an Academy, to disseminate modern knowledge relevant to the military and civilian sectors. His idea was to recruit seasoned instructors from Europe, on the condition that they were not tied to colonial powers in Persia, namely Russia and England. Thus, he sent an envoy of Armenian descent to Europe to recruit instructors from Austria, Prussia, France, and Italy.

Amir Kabir also recruited Mirza-Reza Mohandes-Bashi, who was a graduate of Royal Military Academy in Woolwich, England. Mohandes-Bashi was the best product of Abbas-Mirza's experiment in dispatching students abroad.[5] He was at Woolwich during 1815–1819, and when he returned, became Captain of Artillery, translated several technical and history books into Persian, and participated in the Herat campaign, breaking the fort of Ghurian with modern engineering methods (first Anglo-Persian tug of war over Afghanistan, march of Mohammad Shah on Herat, spring 1838, latter was Abbas-Mirza's son).

Mohandes-Bashi did a superb job in creating a campus environment conducive to learning for this new educational establishment. As a result, was born *Dar-ol-Fonun*, literally House of Crafts, which was inaugurated in November 1851, shortly after Amir Kabir fell to court intrigue.

Nasser-ol-Din Shah took a keen interest in the school; despite his flawed character (a reign of nearly fifty years that began with the murder of his chancellor, with his consent, and ended with his own murder), he was by nature curious and fond of modern inventions and contraptions—tramways, telegrams, rifles, clocks, compasses, cameras, and so on. He had photography equipment brought to Persia from Europe and personally took photographs and annotated them. His son Mozaffar caught the bug and brought motion picture instruments to Persia, kicking off Persia's adventure in cinematography (1900).

Nasser-ol-Din Shah also dabbled in the arts and was a decent sketch artist. He recruited Kamal-ol-Molk, Rembrandt of Persia, into his court, as he spotted him at *Dar-ol-Fonun*, as the new and highly talented art instructor. He often made visits to the school and was for all intents and purposes, the patron of the school. The fact that the school was less than a kilometer away from the royal palace, both enclosed in the walled royal quarter (*Arg*), with the palace having an additional enclosure, facilitated this interaction.

Persia's nobility also threw its support behind this institution, which promised lucrative careers for their offspring as graduates of the Academy, to occupy key posts in government and the military. This was a mixed blessing, as it meant that they monopolized the administration of the school and much of its resources.[6] However, underneath there was a solid conception of how to cover the main branches of learning and to combine classroom work with practical work—in the laboratory, field trips, demonstration projects, surveys, map-making—such that for those who were eager to learn, there was much to absorb. Hence, the school continued its vibrant life until the 1930s, when emergence of other schools of higher learning, most of which had splintered from *Dar-ol-Fonun*, and nearly all of which were absorbed into the University of Tehran, eclipsed the importance of *Dar-ol-Fonun*.

Taghi Erani attended *Dar-ol-Fonun* during 1914–1920. The curriculum consisted of French (language of instruction of European teachers), German, Arabic, Persian literature, history, geography, mathematics, logic, drafting, natural sciences. There was a well-equipped chemistry laboratory built by the Italian Focetti, and textbooks produced by former pupils who had become instructors (Mirza-Kazem Khan Chimie and his son), and a good deal of medical-related subjects. By this time, the military curriculum had been split off into a separate school. In 1918, a College of Medicine was also spun off from *Dar-ol-Fonun*, which shared the same premises but had a parallel administration. Taghi Erani completed his secondary schooling in 1920 as the top student of his class (certificate by Adib-ol-Molk, head of *Dar-ol-Fonun*, attests to this distinction[7]), and afterwards he attended the College of Medicine, completing that program by summer 1922, after which he left for Berlin.

Taghi Erani had a schoolmate at *Dar-ol-Fonun* by the name of Mehdi Azar. He was a year younger than Taghi Erani, and like him had been born in Tabriz, but had grown up in Mash'had. His father had been a deputy in the first *Majles*, representing Tabriz. After the shelling of the *Majles* by the royalists (1908), they had moved to Mash'had, where he completed nine years of schooling. Then, after odd jobs for two years, he travelled to Tehran and managed to enroll at *Dar-ol-Fonun* (program covered grades seven to twelve).

Mehdi Azar overlapped with Taghi Erani during 1920–1922 (Taghi Erani in College of Medicine, Mehdi Azar in high school, on the same site), and like him went on to do the medical program. He graduated from the College of Medicine in 1925. After several years of medical practice in Azerbaijan, he left for France (1928) and became a medical doctor by European standards. He returned to Persia in 1935, progressed in civil administration (the eradication of smallpox his most notable achievement[8]), and had brief encounters with Taghi Erani in mid-1930s. His political memoires have been recorded at Harvard University, as he held a ministerial post in the cabinet of Dr. Mossadegh.[9] He shared his reminiscences of Taghi Erani in two letters to the author, below excerpts:[10]

Taghi Erani was superior to his peers in terms of intellect and mathematics and overall education, and for this reason was subject to their attention and respect. Despite this, he had been reproached by the head of the school, because of certain critiques he had raised about the conduct of the principal.

One day during break-time stroll, he advised me to be prudent in critique of school administration, as otherwise they will create false accusations and cause your isolation amongst the students. And this is indeed what happened.

The principal, who had rejected my application for enrollment in tenth grade, although I had a certificate of completion of the first nine years, would not budge, because I had a threadbare appearance (*maflook*); shoes, hat, attire not commensurate with other students.

So, I appealed to the Ministry of Education, whose head (Dr. Khalil Saqafi) inquired in writing about the reason for the rejection of my application. Since the principal had no credible reason, he had no choice but to take me.

For this reason, he held a grudge against me and one day as he was urging my classmates to learn their lessons well, said for instance this Mehdi Azar, had I not helped him, he would have failed. Outraged by this perjury, with the intent of belittling, I wrote a letter of protest, and this caused his anger against me to further accentuate.

These reminiscences relate to a period when Persia was going through postwar convulsions. The Ottomans and German-backed guerillas had been defeated and the Russians had pulled out of Persia because of the Bolshevik revolution and the ensuing civil war with the white Russians. This left the arena open for appropriation of Persia as a protectorate by the British. Lord Curzon supported by Sir Percy Cox conceived the Anglo-Persian Agreement (August 1919) in complicity with Vosough-ol-Doleh, the Prime Minister (and two of his ministers), which was denounced by Persia's political class and the nation as an act of bondage and triggered a nationalistic backlash (Khiabani uprising in Azerbaijan, May–September 1920). General Ironside was dispatched to Persia (October 1920–February 1921) to assess the situation and organize the withdrawal of British forces stationed in the north (which was source of attack on Lord Curzon in the House of Commons).

General Ironside engineered the decapitation of the Persian Cossack Brigade of its Russian command (sack of General Starosselsky), the appointment of Reza Khan at the helm of PCB, and the mission to march on Tehran without toppling Ahmad Shah (February 1921 *coup*), who increasingly turned into a lame-duck monarch. The *coup* sparked another nationalistic uprising (Pasyan rebellion in Khorasan, April–October 1921), the cementing of the *entente* between Great Britain and Soviet Union over Persia, the withdrawal of Soviet support from the insurgency in Gilan (demise of Kuchak and communist factions, December 1921), and the withdrawal of Soviet and British forces from north Persia. These events led to the consolidation of power by Reza Khan, forced departure of Ahmad Shah under the pretense of medical treatment (November 1923), abolition of Qajar and the birth of Pahlavi (December 1925).

It would not take long for these events to leave their imprint on the life of Taghi Erani; at that juncture though he seems to have been apolitical, at least according to Dr. Mehdi Azar:

> Taghi Erani never spoke of Bolshevism or similar creeds. Perhaps sometimes he spoke of liberty or political regimes prevailing in European countries.[11]

From Taghi Erani's reminiscences about his youth, however, we know that at that time he was filled with nationalistic fervor. He names three poems he composed during this period, two of which are extant, as they were published in the expatriate press in Berlin shortly after his arrival.[12] One of these poems, running twenty-five couplets, uses the imagery of a fallen dame robed in Persia's national flag—emblem of lion holding a saber, sun glowing from its back, against a tricolor background—to convey Persia's dire state of affairs:[13]

> *A dame, visage soothing as the moon, a lion, her companion*
> *Armed with scintillating sword, the lion, amidst halo of the glowing sun*
> *The dame robed in colors of the land, alas now fallen to rugged ground...*

JOURNEY TO BERLIN

We do not know what thought process was involved when Taghi Erani opted to go to Berlin, but we can form certain hypotheses. First, when we look at the track record of the European teachers at *Dar-ol-Fonun*, and this was no doubt a recruiting bias, those who had German as their native or primary language (Prussian, Austrian, Czech of Austrian citizenship, German-speaking Dutch) produced the most impressive results. They pioneered anesthetic surgery in Persia, not long after it became common practice in Europe; drew the first telegraph line from the *Golistan* palace to *Dar-ol-Fonun* and nearby government offices; produced the first modern map of Tehran; measured the height of Mt. Damavand (5600-m); and surveyed the most direct route from the capital to the shores of the Caspian (through mountain passes along Haraz River, obviating the detour via Qazvin), and many other innovations.

Second, the Persians were always in search of that "third force" that would allow them to get some respite from the Anglo-Russian rivalry and complicity over Persia. When they failed to reach alignment with the French (pacts were signed in the Napoleonic era, but nothing came of it), the Persians focused their attention on the newly formed German Empire (1871). But as long as Bismarck was there, Germany paid little attention to Persia, despite Nasser-ol-Din Shah's visits to Berlin in all three of his European voyages (1873, 1878, 1889).

Bismarck was focused on the European balance of power and was not convinced of the economic viability of colonialism at such late stage. (German colonies in east, west, and southwest Africa were acquired in the last few years of his chancellorship in 1880s, by the initiative of the German military, about which he was unenthusiastic.) After Bismarck's dismissal in 1890, things began to change, as Germany considered that it needed stepping-stones in the Near East if it wanted to take on England and her prized possessions, especially India. Thus, the Germans began to make overtures toward Persia though not on the same scale as they did toward the Ottomans. A German school was established in Tehran in 1907, not far from *Dar-ol-Fonun*, though outside the royal quarter (*Arg*). When the Great War broke out, it is no secret that Persia's nationalists, including a host of its prominent poets, became outright Germanophiles. They rooted for the Germans, versifying in praise of the Kaiser, as the mighty eagle took on the bellicose bear and the deceitful fox, both despised in Persia. There was a feeling of affinity toward Germany and the Germans, based on the logic that the enemy of my enemy is my friend.[14]

Aside from these considerations, and the possibility that the Erani family knew of other ethnic Azeris who had settled in Berlin (as indeed there were), there was the plain fact that Germany was leading Europe in science and technology. Its industrial giants in steel, power generation, electronics, and chemical industry were second to none, and Berlin was the avowed Athens of the modern world. Therefore, the lure of Germany and Berlin was overwhelming, despite its defeat in the Great War.

Taghi's challenge was not so much where to go, as he had the credentials and confidence to surmount challenges of an academic nature, but how to pull it off financially, given the bankruptcy of the government (with puny scholarships nibbled at by nepotism), and the estranged relationship between his parents, making financial support unlikely. Taghi did not have a favorable view of his father, and on this occasion the father declined to lend his son any support.[15] The mother gave her son as much support as she could, materially and morally, and the rest he managed through his own initiative. Additionally, moonlighting while pursuing his studies produced exceptional opportunities for his life experience and intellectual growth.

When Taghi embarked on the journey to Berlin in August 1922, he was a young man of twenty. He was no doubt filled with emotions as he bid farewell to his mother, whom he venerated; and his sisters, thirteen, eleven, and two years of age. His father would be straddling between Tabriz and Tehran, while Mash'hadi-Reza Damavandi would be attending to their chores. He would have to leave from Qazvin Gate on the westside of the citadel, the same he had taken when he arrived in Tehran as a ten-year-old boy. On horse-drawn *delijan*, he would head to Qazvin, and instead of continuing westbound toward Tabriz, a turnoff to the north would take him to the city of Rasht in Gilan province, a few kilometers from Port Enzeli. Then, he would take the steamer on the Caspian, with stopovers in coastal towns of his ancestral Eran—Astara, Lankaran, shores of Shirvan, Baku (behind which Ganja was tucked away)—landscape on which he would be setting eyes for the first time, and ruminating about the distant past. Then, on to large-gauge trains from Baku through Soviet Azerbaijan, *Russland*, and Polish territories, canvassing the still scarred landscape of the Eastern Front, onto *Deutschland* and Berlin, where new horizons awaited him in a metropolis that was groping toward its Age of Pericles.

NOTES

1. See Rosen, *Oriental Memories of a German Diplomatist*, p. 77.

2. My description of old Tehran is based on the five-volume work of Shahri, the article of de Planhol, the first modern map of Tehran by Krziz (1857), and the map of Wagner and Debes (1914).

3. For Erani family attendant, see Boroujerdi, *Erani Faratar az Marx*, pp. 459–460; this was his interrogation by the police shortly after the arrest of Taghi Erani. He says he is forty-one years old, married, has three children, lives in Tehran (gives address), and brings fresh clothing and food for Dr. Taghi Erani in prison, prepared by Taghi Erani's mother.

4. See Kasravi, *Qiyam'e Sheikh Mohammad Khiabani*, pp. 91–92: "From his teachings of Ptolemaic astronomy [referring to Khiabani], it may be understood that at that time [1906] the old ideas were still dominant in him, as in those days Copernican astronomy was to some extent taught in Tabriz, by few people."

5. See Hashemian, *Tahavolat'e Farhangi'e Iran*, pp. 35–36, on the career of Mohandes-Bashi.

6. See Hashemian, *Tahavolat'e Farhangi'e Iran*, pp. 150–151, on Ehtesham-ol-Saltaneh's critique of Qajar aristocrats who abused the resources of *Dar-ol-Fonun*.

7. For Erani's distinction as top student of *Dar-ol-Fonun*, see facsimile produced in Chaqueri, *Taghi Erani dar Ayeneye Tarikh*, p. 202.

8. See Azizi, *A Brief History of Smallpox Eradication in Iran*, crediting Dr. Mehdi Azar and three others to have accomplished this feat.

9. For the political biography of Dr. Mehdi Azar (1903–1993), see Zia Sadeghi, *Mosahebe ba Dr. Mehdi Azar*.

10. Private correspondence with Dr. Mehdi Azar.

11. Ibid.

12. Erani poems are *Madar'e Meehan*, *Molood'e Hazrat'e Rasul*, and *Takht'e Jamshid*; latter is not extant and seems to have been a collection as he refers to it in the plural; see Boroujerdi, *Erani Faratar az Marx*.

13. Poem *Madar'e Meehan* is reproduced in Boroujerdi, *Erani Faratar az Marx*.

14. See Aryanpour, *Az Saba ta Nima*, pp. 317–332.

15. For Erani's view of his father, see Momeni, *Parvandeh*, p. 283: "During secondary schooling at *Dar-ol-Fonun*, in spite of having a father who was affluent and with means, and notably when abroad, I lived frugally because of the irresponsibility of the father." Also in Khame'i, *Panjah Nafar va Se Nafar*, p. 16: "My father was a real Casanova."

CHAPTER 4

Berlin (1922–1928)

German prowess in science and technology lured Taghi Erani to Berlin, but practical necessities exposed him to a parallel universe, which was to have a decisive effect on his life. This unintended track was the humanities. It began as a menial job in a Persian publishing house in Berlin to secure temporary lodging and a trickling of income. This outfit became his moonlighting den, while in the daytime he attended the university to pursue his ambitions in science.

Gradually, the moonlighting took a serious turn with the annotation of some of the classics of Persian literature, the chance encounter with Friedrich Rosen, the lectureship at the Oriental Languages Institute, the tome on psychology, and political activism as the culmination of psychological introspection.

Unlike Erani's pre-Berlin years, where our knowledge of his life is scant and confined to family genealogy, a few photographs, school records and an account from a schoolmate, the years in Berlin are well documented. We have the correspondence in the Weimar archives,[1] a trickling from Soviet archives,[2] the doctoral dissertation,[3] and publications authored by him in Berlin.[4] These consist of ten articles in the expatriate press, three book annotations in the field of orientalism, an *oeuvre* on psychology, a prologue to his *Series in Exact Sciences* and its first installment, a textbook on physics. As well, there is testimony by a contemporary in Berlin, a literary figure who passed away in 1997 at the age of one hundred and five (Jamalzadeh), and photographs in the family archives. These, with the wealth of literature available about the

© The Author(s) 2019
Y. Jalali, *Taghi Erani, a Polymath in Interwar Berlin*,
https://doi.org/10.1007/978-3-319-97837-6_4

history, science, culture, and politics of this period—the Weimer era—are sufficient to develop a vivid picture of this phase of his life.

NOTABLE ENCOUNTERS, ADDRESS TRAIL

Name-dropping may be the shortest route to give a flavor of this season of his life—personalities that he came to know firsthand or with whom he came to have tangential interactions. This is at a place and period in modern history which was in some sense the apotheosis of scientific, artistic, and intellectual brilliance, but fraught with contradictions that at the end produced a spectacular apocalypse, the Third Reich.

Leibnizstrasse 43 in the leafy suburb of Charlottenburg in west Berlin is where Taghi Erani stumbled on Friedrich Rosen, the retired states-man who in troubled times found solace in the Quatrains of Omar Khayyam—*gently sip the wine and strike the harp, for of those who are here none will remain, and of those who have gone, none will return*. Rosen was visiting the Persian print-shop to get his latest work out on the sage of Neyshabur; this was in spring 1924.

Only months after Rosen's resignation as Foreign Minister (October 1921), his successor Walther Rathenau—these two being the most prom-inent Jewish statesmen of the Weimar era—was assassinated by right-wing extremists (June 1922), for a political stance that Rosen as well had upheld, compromise with the victors of the Great War on the issue of reparations. For Rosen, the assassination came as a grave tragedy and a near miss, the London Ultimatum which triggered his resignation, a blessing in disguise. This hurled him towards his lifelong passion, orien-talism, notably Khayyam, and a chance encounter with Taghi Erani.

Through Rosen, Taghi Erani came to know prominent figures of German orientalism: Eugen Mittwoch, master of Amharic (Semitic lan-guage spoken in Ethiopia) and a notable Hebraist, head of the Oriental Languages Institute at the University of Berlin, Erani's prospective employer, and unbeknownst to him, a member of German intelli-gence (and later the British intelligence, as the Nazis banished those of Jewish origin, Mittwoch switching sides); Carl Heinrich Becker, the Minister of Culture of Prussia, a noted orientalist (Arabist), surpassing Mittwoch in professional reputation, whose archives contain corre-spondence from Taghi Erani;[5] Wilhelm Litten, a diplomat and Persia scholar, author of several works on contemporary Persia (an oddity among orientalists, who much preferred the "elevated" ancient world to the fallen present). Litten has a detailed description of his interview with Taghi Erani, conducted at the Oriental Languages Institute, at *Dorotheenstrasse 7*.[6]

Taghi Erani became a lecturer at this institute during 1925–1928, teaching a weekly seminar (Monday and Thursday evenings, seven to eight) on the theme of Oriental Rhetoric, Style and Logic, drawing on Persian, Arabic, and Turkish sources, a course in comparative rhetoric delivered in German.[7]

***

Leibnizstrasse also featured a bookshop[8] which was one of the front businesses of Willi Münzenberg's operations in Berlin, the Western European Bureau (WEB). This was a front for the Comintern[9] tasked with galvanizing support for the Soviet Union among opinion-makers: authors, playwrights, filmmakers, celebrity scientists, and steering the left-wing political parties, notably the mass parties (German, French, Italian communist parties) through financial inducement, intelligence, logistics, coercion and terror.

Willi was focused on propaganda, became a media mogul, and was rather autonomous, although he relied on Soviet handouts to run his operations. He knew Lenin personally from his pre-revolution days in Switzerland. He managed to eke out a lucrative lifestyle in posh Berlin (Tiergarten, *Zeltenstrasse 9a* his residence, "Red Millionaire" his alias).[10] He made it as Reichstag deputy, was a formidable orator, and played on multiple boards, on each in multiple dimensions.

One of his publicity stunts, during the year to celebrate the tenth anniversary of the Bolshevik revolution, was the Brussels conference on the theme of anti-colonialism, to which he was able to draw such third-world celebrities as Chiang Kai-shek (two months before he turned on his communist allies in China, and broke with Moscow), Nehru of India, Hatta of Indonesia, a leading figure in Indonesia's independence from the Dutch (later Prime Minister under Sukarno), and Sen Katayama the communist leader from Japan, representing a standpoint opposed to Japanese annexationist designs on Asia.[11]

To this, we must add the pacifists—Albert Einstein, the renowned physicist, Romain Rolland, the French dramatist, Diego Rivera, the Mexican muralist and painter, and Taghi Erani, his neighbor from *Leibnizstrasse*. Erani, who was someone with a modest portfolio, a leader of an anti-colonial organization focused on Persia, with a founding declaration and a four-point charter submitted to this gathering, was nevertheless a minor figure in this conference.[12] A letter in Comintern archives confirms the connection between Taghi Erani and Willi Münzenberg.[13] There is also a photograph from this conference showing Taghi Erani in attendance.[14] The conference was held at Egmont Palace during February 10–15, 1927, with 174 delegates representing 104 organizations.

Willi was crafty in conceptualizing the mediatized effect of any undertaking he conceived, and he was both despised and admired by Goebbels, Hitler's propaganda chief. He showed great courage to counter the Nazis even after they assumed power (January 1933), though that window was extremely short, and he had to flee to France after the Reichstag fire (February 1933).

Taghi Erani also despised the Hitler apparatus and had much to say about its ringmasters based on his personal observations in Berlin, during autumn 1934 and summer 1935:

> Meitner, a lady in her forties, who until recently was a professor at the University of Berlin, and Head of Atomic Research at Dahlem laboratories [where Erani's lab was located], was stripped of her first role due to her Jewish background.[15]
>
> [She lost her professorial job due to its high visibility; what saved her from losing her second job was that she was Austrian, so not subject to the civil service law that barred Germans of Jewish heritage from state positions; when Austria was annexed by Germany, that changed; she left for England prior to this, via Netherlands and Sweden.]
>
> Only an imbecile can believe a Turk has less aptitude than a Bulgarian [being Asian not European], or a Jew has less aptitude than a German… Rosenberg who must fabricate racial theories for the fascists would do well to venture to neighborhoods in north Berlin, where most likely he has never set foot. The German who, according to him, is above all races is crushed by poverty, misery, refuge to alcohol, mass internment in suffocating shelter, work in toxic space and joblessness.[16]
>
> After Hitler grabbed the rein, and the passage of several months, the masses expected fulfillment of the promises made to them, to the point that rumblings began about a second revolution. But as the clique in charge was unable to fulfil its promises (such as closure of large department stores to curb unemployment) and saw these demands as a threat to its rule, it resorted to terror. Last summer [1934] the head of the *Stormtroopers* and his allies were all executed, as they felt the discontent

with acuity. The reorganization of the *Stormtroopers* is the continuation of the same policy. *Agence Reuters* reports: 'Lutze attempting to form new *Stormtroopers* that completely back the National-Socialist regime and collaborate with political leaders of the party.'[17]

On 16th July [1935] a pogrom took place against the Jews on German soil; these are reminiscent of acts that took place in Russia in bygone era, a political show stemming from economic failure.[18]

Taghi Erani's observations are reminiscent of Joseph Roth's *feuilletons* from Nazi Berlin during the same period. He had a good read on the Third Reich when and as it was unfolding, largely due to his association with prominent political figures in the 1920s.

While in Berlin in 1920s, Taghi Erani had dealings with Rudolf Breitscheid, spokesman of the independent social-democratic party in the Reichstag (USPD). He made an appeal to Breitscheid to not allow the government of Persia to meddle in basic rights in Germany, such as freedom of expression. Persia's embassy was pressing the government of Hermann Müller to expel a student engaged in anti-Pahlavi activism (Ahmad Asadi). The British pressed for the same, claiming that literature from a certain group to which this student belonged had been dispatched to India, inciting anti-British sentiments.[19] What no one knew is that this student was a collaborator of Taghi Erani in a political group they had founded (along with three others) in the wake of coronation of Reza Shah (April 1926), at the culmination of his ascension to power to which the British had given much support. This group had links with the German left, inviting one of its veterans to their inaugural meeting (Georg Ledebour of the USPD, one of the leaders of the revolution of 1918–1919).[20] Taghi Erani, however, made his plea on behalf of the association representing Persian students in Germany, of which he was the secretary. Breitscheid decided to lend his support by publicizing this case in the leading national daily, *Vorwärts* (December 19, 1928, *Die Republic lässt lich missbrauchen, Liebesdienste für den Schah—Republic Abused, Love Affair with the Shah*), and scoring a point against Chancellor Müller. He asked for an inquiry as the Foreign Ministry seemed to be manipulated by a despotic regime, using trade as blackmail. This initiative, however, was unsuccessful and led to the threat of expulsion brought against Taghi Erani. This strand of the story marks his debut in political activism, leading to his deeper reflections and radicalization.

***

Fasanenstrasse 22, Charlottenburg, was Taghi Erani's last address in Berlin. The name means Pheasant Road, and it was in a posh neighborhood, as by then he had begun to receive financial support from the government of Persia, though he shared this flat with a fellow student, a compatriate.[21] A plaque on this building with municipal seal commemorates this "anti-fascist" intellectual from Persia.[22] Taghi Erani had moved to this building in 1926. In a letter to Carl H. Becker, Prussia's Minister of Science-Art-Education, dated October 22, 1926, he gives this address, while in the December 1925 directory of university lecturers, his address is *Bl-Charlottenburg, Weimarerstrasse 27*.[23]

Erani lived at *Fasanenstrasse* at least two years, possibly three, as he left Berlin in December 1928, shortly after completing his doctoral dissertation and the breakout of the expulsion threat. In the university directory of December 1928, the *Fasanenstrasse* address is given, as well as the title Doctor of Philosophy,[24] and he did not return to Berlin until autumn 1934 and summer 1935, when much had changed, as alluded to in his observations of Third Reich Berlin.

That this last address was in a posh neighborhood there is no doubt. The celebrated chemist Fritz Haber, the father of nitrogen fixation for industrial production of ammonia (for which he received six-figure royalties from BASF each year[25]) and the father of chemical warfare (lethal gases used in the war[26]), bought a mansion on *Fasanenstrasse* in 1927, for his second wife, Charlotte, from whom he was separating.[27] Haber lived in the suburb of Dahlem, 5-km south of Charlottenburg, where his research laboratory was located. *Fasanenstrasse* was chosen so his two children could visit him with ease. And Haber's son, Hermann, from the first wife, was a contemporary of Taghi Erani at the Chemistry Department at the University of Berlin.[28] Hermann (b. 1902) obtained his doctorate under the colloid chemist Freundlich, who appears in Erani's *Lebenslauf*, bio produced for his doctoral dissertation, where he lists professors whose lectures and exercise sessions he attended (*Vorlesungen und Uebungen*).[29]

Taghi Erani studied organic chemistry, and his doctoral research was supervised by Professors Hermann Thoms and Carl Mannich, the latter with a class of chemical reactions to his name.[30] And Thoms was the founder of Institute of Pharmacy at the University of Berlin (also at Dahlem, founded 1903).[31] Though he was an organic chemist, his goal was to create a multidisciplinary approach to the study of drugs. Erani's background in medicine (he had done a two-year program in Persia) was a good fit for this approach. The American Pharmaceutical Association made Thoms an honorary member in 1923, though the wounds of war were still fresh (highlighted in the award, stating it is high time to move on, notably for such an outstanding scientific figure), for Thoms' pioneering role in the field of pharmaceutical chemistry.[32]

German chemistry occupied much of Erani's interest. We will learn of its existential dimension, as well as Erani's research work and his commentary on the architects of modern science (Nernst, Planck, Einstein, all of whom appear in his *Lebenslauf*, along with other notables, Bodenstein, Pringsheim, Rosenmund, eighteen in all), including commentary on leading women scientists, Lise Meitner and Irène Curie, whose work on induced radioactivity he followed with great interest (what became the basis for engineering the bomb).

In his commentaries, Taghi Erani describes not only the science, but also the scientists, notably Planck, Einstein, and Nernst, personalities that were badly bruised in those caustic times. Nernst had lost both his sons in the war,[33] and Planck lost his eldest in the war (Battle of Verdun) and later lost the youngest to the Third Reich, during the foiled attempt to assassinate the Führer.[34] Einstein escaped unscathed, if we gloss over the unease he felt for having lobbied for the atomic bomb, abandoning his pacifist philosophy.[35] The fate of others was not much better—Meitner had to flee to the Netherlands, then Sweden and England, when anti-Semitism reached its zenith,[36] and Hermann Haber who was in France when the Nazis came to power, had to flee when the French threw in their lot with Hitler. He made it to the USA, but committed suicide shortly after the war.[37]

***

Erani's *Lebenslauf* also contains the name of Professor Koehler. According to Professor Wertheimer,[38] historian of psychology, it is nearly certain that this is Wolfgang Köhler. Köhler had studied acoustic phenomena and perception under Planck and Stumpf, was a leading proponent of *Gestalt* psychology, laying emphasis on the holistic nature of perception, and was the director of the Psychology Institute at the University of Berlin (1921–1935). The institute was housed at *Schlossplatz*, former residence of the Kaiser. Psychology was another field to which Erani was drawn while in Berlin, where he produced his *magnum opus* in December 1927. His work in orientalism had exposed him to the evolution of thought in the East and West, the history of ideas, the development of magic, mythology, religion, philosophy, and science. This brought him in touch with Wilhelm Wundt's work on *Völkerpsychologie* (cultural psychology). And he made an effort to connect psychology from the vantage point of the individual and neurobiology where his intuition from medicine led him to its aggregate form at cultural level shaped by historical experience and memory. This strand of his intellectual curiosity is reflective of his journey in self-reflection and gradually led to his metamorphosis and turn to political activism and radicalism.

NOTES

1. Erani in German archives; my findings are listed under *Bankverbindung, Verzeichnis-Personal,* and *Verzeichnis-Vorlesungen.* There is also previously published or referenced archival material in Mahrad, *Die deutsch-persischen Beziehungen von 1918–1933,* in Ahmadi, *Tarikhche Ferq'e Jomhooriye Enghelabi'e Iran,* in Boroujerdi, *Erani Faratar az Marx,* and in Chaqueri, *Taghi Erani dar Ayeneye Tarikh.*
2. Erani in Soviet archives; see Chaqueri, *Taghi Erani dar Ayeneye Tarikh,* pp. 87–126, 158–179, 254–259, and 263.
3. See *Erani, Tagi, Die reduzierenden Wirkungen der unterphosphorigen Säure auf organische Verbindungen.*
4. The articles, ten in total (number noted in parentheses) appeared in the following journals—*Iranschähr* (4), *Farangestan* (1), *Azadiye Shargh* (2), *Sanaye'e Alman va Shargh* (1), and *Elm va Honar* (2). See bibliography.
5. *Bankverbindung* contains Becker letters which are replies to Erani letters; discussed in Chapter 7.
6. Litten interview of Erani, conducted on January 14, 1925; summarized by Litten and submitted to Mittwoch in a three-page memo dated January 16, 1925. See *Bankverbindung,* pp. 56 (recto-verso) and 57 (front-face).
7. See *Bankverbindung,* p. 149 (recto-verso); also see *Verzeichnis-Vorlesungen* (*Wintersemester 1925/1926 zu Wintersemester 1928/1929*).
8. See McMeekin, *The Red Millionaire,* p. 87.
9. Comintern was formed in 1919, dissolved in 1943, also known as the Third International (first being the one in which Marx was active, second the one that disintegrated during WWI), vehicle used by the Soviet Union to steer member parties toward its agenda of low-level warfare with adversaries. It was formed in the wake of the Russian revolution in the midst of civil war as self-preservation mechanism; it was dissolved

after the Allies lent their support to the Soviet Union to defeat Nazi Germany. See Hallas, *The Comintern*, for a brief history.

10. See McMeekin, *The Red Millionaire*, p. 204.
11. Ibid., pp. 196–201.
12. See Chaqueri, *Taghi Erani dar Ayeneye Tarikh*, p. 179, for the four-point charter.
13. CRRP letter dated March 29, 1926 addressed to Comintern Bureau of East, signed by Erani, stating they can be contacted via Münzenberg or Gibarti; see Chaqueri, *Taghi Erani dar Ayeneye Tarikh*, pp. 158–159.
14. Erani photograph, Brussels Conf.: http://historian.blog.se/tag/virendranath-chattophadyaya/.
15. See Erani, *Donya*, 12-issue Volume, p. 269.
16. Ibid., p. 329.
17. Ibid., p. 292.
18. Ibid., p. 386.
19. The student subject to expulsion was variously known as—Asadi (Azadi in German), Azad-zadeh, Azadov, and Darab (pseudonym). See Mahrad, *Die deutsch-persischen Beziehungen von 1918–1933*, pp. 331–337. See Ahmadi, *Tarikhche Ferq'e Jomhooriye Enghelabi'e Iran*, pp. 24–25, for Erani's plea to Breitscheid (*Aufzeichnung des AA, V. 23, November 1928 zu III 06273/28*).
20. See Chaqueri, *Taghi Erani dar Ayeneye Tarikh*, p. 171. Ledebour had been member of the "Provisional Revolutionary Committee" during the 1918–1919 revolution; see Haffner, *Failure of a Revolution*, p. 132.
21. Taghi Erani shared this apartment with his close friend, Ahmad Emami, who studied medicine. Emami's name and address is given in the German police report, investigated at the request of Persia's embassy in Berlin; but he was found to have no political involvements. German police memo is given in Boroujerdi, *Erani Faratar az Marx*.
22. See *Gedenktafel-Tagi-Erani*.
23. See *Bankverbindung*, p. 304.
24. See *Verzeichnis-Personal* (*Dezember 31, 1928*).
25. See Stoltzenberg, *Fritz Haber*, pp. 95–96.
26. Ibid., pp. 134–135, 144.
27. Ibid., p. 188.
28. Ibid., p. 179.
29. See *Lebenslauf* (bio) in *Erani, Tagi, Die reduzierenden Wirkungen der unterphosphorigen Säure auf organische Verbindungen*.
30. A class of chemical reactions used in the organic synthesis of polymers, detergents, plant growth regulators, pharmaceutical products (antibiotics, antidepressants), and building blocks of certain biomolecules (peptides, nucleotides).

31. See Kremers, *Hermann Thoms*, on the career of Professor Hermann Thoms (1859–1931), leading figure in German pharmaceutical chemistry, and peerage awarded to him by the American Pharmaceutical Association.

32. Ibid.

33. See Coffey, *Cathedrals of Science*, p. 110.

34. See Planck, *Scientific Autobiography*, pp. 7–8; also Coffey, *Cathedrals of Science*, p. 317.

35. See Seelig, *Albert Einstein*, p. 115, for his speech of December 10, 1945; also see Long, *Einstein and the Atomic Bomb*.

36. See Nye, *Before Big Science*, p. 200.

37. See Stoltzenberg, *Fritz Haber*, pp. 176–183.

38. Private correspondence with Prof. Michael Wertheimer.

CHAPTER 5

Foray in German Science

Before 1945, when the atomic bomb demonstrated the importance of physics to military might, it was not physics but chemistry which mesmerized the elite of nation-states eager to join the leading nations of the world. And the elite were not political figures only, but also financial, industrial, scientific, and cultural personalities. This is certainly true of Japan during Meiji, "Prussia of the East," but here our focus is on the west, Prussia plus the other German-speaking states they came to amend, creating the German Empire (1871).

The year 1888 might be taken as a turning point in the history of the German Empire, the year of three emperors, the grandfather and father passing away in quick succession, leaving the grandson at the helm. The year 1890 was the decisive turning point, when Bismarck was sacked by the young emperor, who was determined to defend Germany's growing commercial interests with a navy capable of holding its own against the greatest naval power of the time, Great Britain.

This assertive policy required a great many things, for war was always on the horizon, and the sentiment of encirclement engendered by the *Triple Entente* (England, France, Russia) was felt with great acuity. One need only glance through the Kaiser's memoires to sense the gravity of the situation.[1]

A determined approach to the promotion of science was an important pillar of this strategy, and in the sciences, there was none more primordial than chemistry. Physics had been making strides, though its greatest days were still far into the future, and it was in some sense the science for

© The Author(s) 2019
Y. Jalali, *Taghi Erani, a Polymath in Interwar Berlin*,
https://doi.org/10.1007/978-3-319-97837-6_5

the mystics and the poets, who cherished uncovering the mysteries of the universe in their blissful solitude, with ever greater degrees of mathematical abstraction. But chemistry was for the worldly creature, for it was a most pragmatic science to begin with, requiring hordes of researchers, and even the most theoretically inclined chemist was after uncovering truths that would have untold practical benefits.

The hundredth anniversary of the founding of the University of Berlin was not a date to be missed in which to make public this commitment to science, proving that not only there was a vision guiding the nation into the future, but that there was unanimity of purpose amongst the architects and artisans of the nation-state.

The centennial would be in 1910, but it was on January 11, 1911 that the pompous ceremony for this occasion took place, in the presence of the Kaiser and the most eminent scientists, industrialists, financiers, and cultural icons of the up-and-coming Germany. It was precisely at this point, as well, that the many paradoxes with which this enterprise was fraught began to unravel, one after the other.

VITAL SCIENCE FOR GERMANY, PREWAR AMBIENCE

That chemistry was the vital science for Germany, there could be no questioning—where would Germany be without its chemical manure to boost agricultural productivity and feed its growing populace? The very realization that nitrates played a key role in plant growth and that these could be artificially supplied with the same effect as natural manure was a German discovery, heralded by Liebig, the prophet of organic chemistry.

If Germany did not have enough of the salts rich in nitrates, as was the case, making it dependent on imports from Latin America (Chile), a shaky business for an encircled nation, then it could fixate the free nitrogen in the air to produce ammonia—that is, if it had deep insight into the workings of chemical reactions (including reactions of the infinitesimally reactive elements), and what combination of catalyst, pressure and temperature would provide a yield that would make it an industrially viable process. Indeed, the Germans did not have one, but three such methods, only that one was far better than the others. This was the famous Haber process, for which he won the Nobel Prize in Chemistry in 1918, industrial ammonia having been produced by his method as early as 1913.

Nitrates were not only important for agriculture but also important for armaments, as in explosives, with which the German high command was obsessed, though their fixation faded when they ran over Belgium with its ample stock of mines and minerals. However, they maintained their enthusiasm for any weaponry that could break the gridlock on the Western Front. Unlike its adversaries, Germany had neither the demographic nor the geographic resources to get engaged in a drawn-out conflict, and this is all that was on the horizon with the war of the trenches.

Chemistry could break this deadlock, for if a lethal gas could be engineered with greater density than air, once delivered by barrage of artillery shells encapsulating the gas, it would naturally sink into ground level and wipe out its targets.

To be fair, it was not only the Germans who embarked on this path, but also the British and the French,[2] and in all this, they were united in one sense—that chemistry potentially held the key to success. The Germans on their part continued on this path even after the war, by offshoring the manufacture of chemical weapons in Russia and Spain (Morocco), with a buyback option, for there was Allied scrutiny at home.[3] To prove chemistry could help in peacetime beyond the camouflaged manufacture of lethal gases, Haber's team even envisaged extraction of gold from seawater to help pay for the punitive reparations demanded by the Allies, notably the French, but this proved technically infeasible.

German chemistry was not only about war-preparedness and military application. Above all, it was about economic supremacy and capturing world markets. What better way to achieve this than by making all major economic activity inextricably dependent upon it—mining (chemicals for explosives, extraction, purification), power generation (chemical process for gasification of coal, hydrogenation, to produce the most efficient fuel for generation of electricity), steelmaking (production of carbon-rich coke from coal and petroleum, as solid fuel to make steel), textiles (bleaches, synthetic dyes and fabrics), construction materials, agriculture, food processing, beverage and brewing industry, petroleum processing, gasworks, glassworks, medicine, sewage and water treatment, and the list goes on, even for the turn-of-the-century chemical industry, modest by today's standards. It was only a question of who would have the lion's share of this industry of industries—the British, the French, the Germans, or the Americans.

The British had pioneered much of chemical science and industry (Priestley, Davy, Dalton, Faraday, Perkin to name a few), but due to their distraction with the colonies, they were no longer in a leading position. The German-born Prince Albert, the husband of Queen Victoria, did much to promote the sciences in Great Britain and its marriage to industry (hence the Prince Albert Prize), but this focus was lost with his premature passing away. The French were at war with themselves, and the father of modern chemistry, Antoine Lavoisier, was guillotined in the mayhem of the revolution. The French seemed satiated as an agricultural powerhouse to want to make a run for leadership in industry.

The Germans, however, were helped by their so-called handicap. The multiplicity of hitherto autonomous German states meant that each had its own patrons of science, and depending on the local realties—from the agricultural east (east of Elbe river), to the industrial west (the Rhine region), to the minefields of the Ruhr valley, the plains of the north and the highlands of the south, dotted with urban agglomerations—supported the local technical schools, universities, and institutes. Local industries, estates, and businesses needed skilled workers, technicians, engineers, and scientists to run their operations. In the universities, it had become the German way to make the laboratory the centerpiece of the classroom (developed to an art form by Liebig and his students, Kekulé, Hofmann), and to make independent research a precondition for academic recognition and advancement.

Further, coursework and research were portable from one institute to the other across the German states. Students could study multiple disciplines (chemistry, agriculture, medicine, pharmacology), and this was not only possible but encouraged. No wonder many flocked to Germany to partake in this revolutionary approach to education, especially in the sciences and chemistry, including Americans, the most accomplished of whom received solid German education (Richards, Harkins, Lewis, Langmuir).

The Germans also sent missions to America for facts-finding (one led by the omnipresent Haber) and soon realized that the real threat came not from their European neighbors, but from the Americans. This was not only because America was thirty times larger than Germany and endowed with great natural resources, but also because it had embraced the gospel of capitalism, and in chemistry it had adopted the German way—the laboratory, the independent research, the cross-disciplinary focus, the tight liaison with industry, the backing of finance. Thus, unless Germany stayed vigilant, it was bound to be overtaken if not overthrown by the USA.

This was the subtext of the keynote speech delivered by the legendary chemist, Emil Fischer, at the Centennial Ceremony of the University of Berlin on January 11, 1911.[4] Only that he was neither alarmist nor evangelical; it was not his style and the occasion did not call for it. The gathering was to celebrate the founding of not one but two cutting-edge institutes of chemical research, in the capital of a nation determined to shape its own future, in symbiotic relationship with the University of Berlin, in which a great deal of chemical research was already underway, not to speak of the independent technical and vocational colleges, which also thrived in this ecosystem.

In his keynote, Fischer gave a wonderful resumé of German progress in chemistry over the preceding decades. Until then fifty percent of Nobel laureates in Chemistry were German, and even twenty years on this proportion had dropped only to forty percent. With specimens and exhibits, Fischer imparted the importance of working on boundary fields (including his latest focus, protein synthesis), the importance of using precision instruments, of not burdening the researcher with teaching duties, creating a solemn environment for fundamental thinking to grasp the truths of nature, and solving the most pressing problems of humanity, and in the process, of the nation.

In some sense, it was the philosophy that scientific progress is the essence of nationalism, and nationalism a stepping-stone to globalism and humanism. It is here that the cracks begin to appear—the scientist is obsessed by progress, the industrialist by products, the financier by profits, the bureaucrat by power, the politician by prestige and posterity.

Before he passed away, or rather before he took his own life by chemical poisoning to end the agony of the loss of his first son in the war and the second to suicide (triggered by a compulsory draft to the front), and the shattering of the world around him, Fischer finally broke his Stoic reserve:

Modern warfare is in every respect so horrifying that sensible people can only regret that it draws its means from the progress of the sciences. I hope that the present war will teach the peoples of Europe a lasting lesson and bring the friends of peace into power. Otherwise the present ruling classes will really deserve to be swept away by socialism.[5]

Such were Emil Fischer's words as the war was drawing to a close.

POSTWAR AMBIENCE, TAGHI ERANI AND CHEMISTRY

The postwar scientific ambience was the world in which Taghi Erani was immersed. On the whole, there was a rise in the culture of unreason, as the catastrophe was attributed to blind submission to rational thinking and mechanical progress.

Fichte, Schopenhauer, Nietzsche rose from the dead; Spirit, Will, Intuition became the currencies of choice, the only wares of any worth. Rathenau preached the gospel of *New Society* ("Jesus in Tuxedo" is how he was mocked by detractors, as he had a spiritualistic message, but as a wealthy industrialist, head of the AEG, was always in a refined setting and in impeccable attire), and Bergson propounded a lofty form of mysticism in the garb of ethics and philosophy.

The scientists caught this mood as well. The philosophy of positivism (Auguste Comte) marked by extreme caution to make generalizations, gained currency. It was deemed safer to forgo grand schemes for the certainty of the narrow and the minute. Einstein propounded the possibility of the finiteness of the universe, as it yielded a finite mass density and the certainty that everything could hang together by gravity. Later he abandoned this view, which hinged on the notion of a static universe with a radius that would take light only ten billion years to traverse.

Meanwhile, steady progress continued to be made at the boundary of chemistry and physics—Röntgen's X-ray, Becquerel's radioactivity, Thomson's electron, Planck's quanta, Rutherford and Bohr's atomic model, chemical bond as the exchange or the sharing of electrons and ionization as the separation of elements with exchanged electrons (Lewis). As well, induced decay of heavier to lighter elements by irradiation (Lithium to Helium, Cockcroft and Walton) and conversion of lighter to heavier elements by irradiation (Boron to Nitrogen, Magnesium to Silicon, Aluminum to Phosphorus, Irène Curie) were discovered. Furthermore, there was the discovery of the neutron (particle of the atomic nucleus separable into proton, electron, and extra radiation, Chadwick), nuclear breakup by bombardment of the heaviest known element, uranium, with neutrons (Hahn-Meitner-Strassmann, aiming to produce elements heavier than uranium, but triggering instead the fragmentation of uranium). Nuclear breakup by the same process was discovered, with a focus on slow neutrons causing release of yet other neutrons in a chain reaction (Fermi), the latter leading to weaponization (Manhattan project).

In some sense, it was the progressive moving of physics into center stage in a space of fifty years (1895–1945), the emergence of quantum mechanics, while shedding light on chemical phenomena and producing an array of boundary disciplines (radiochemistry, quantum chemistry, electrochemistry, physical chemistry, thermodynamics), as chemistry proper gravitated toward macromolecules of the living world, towards yet another boundary.

As far as the aftermath of the Great War is concerned, there was an East–West divide in terms of the prevailing sentiment towards the future. While scientific communities in the West tended to act with circumspection, the aspiring scientists and intellectuals of the colonial world who were flocking to Europe were on the whole upbeat about the future.

The Great War had caused a weakening of colonial powers, including the victors. The colonies were making strides to gain their independence, China and India leading the pack. Instead of feeling shipwrecked, it was time to devour the fruits of the West's scientific and technological progress. How to escape the disease inflicting the West, the compulsion to wage war to protect economic interests, fueled by high science and unleashed at the scale of the globe, is something that remained fuzzy and had to be worked out down the line.

It is in this spirit that Taghi Erani seems to have taken in the ambience of the period. In August 1922, he had received his credentials from *Dar-ol-Fonun* and the College of Medicine in Tehran, and by September of that year he had arrived in Berlin. Winter semester began in October, which was his debut at the *Friedrich-Wilhelms-Universität zu Berlin*. He was initially admitted as a guest student (*Gasthörer*, semester running from October 16, 1922 to March 15, 1923). Then, he obtained admission to the *Philosophischen Fakultät*. This naming convention was a vestige of the past. The Faculty of Theology was the established branch, and anything smacking of science was placed in the Faculty of Philosophy. Medicine was one of the subdivisions of philosophy, and chemistry one of the subdivisions of medicine. However, with the growth of chemistry in the nineteenth century, it had become an autonomous discipline.

Taghi Erani spent eleven semesters at the university, the first as guest student and ten in the Faculty of Philosophy.[6] His first semester in the Faculty of Philosophy ran from April 16, 1923 to August 15, 1923. This was the summer semester, and the dating for summer and winter semesters repeated yearly, with no change.

Two examinations were required to ensure that students had acquired adequate knowledge of inorganic, organic and physical chemistry, as well as physics, at both theoretical and experimental levels. This is what ensured that a student could move from one university to the other. Examinations were administered by an independent body (*Verband der Direcktoren selbständiger Unterricht-Institute für Chemie an deutschen Universitäten*).[7]

Taghi Erani passed his first examination in March 1925 (*Verbandsexamen*), and his second in March 1926 (*Doktorandumprüfung*). In April 1926, he began his doctoral research (*Doktorarbeit*) and he completed this research in March 1928. In May 1928, he requested that his doctoral defense be postponed, due to illness. In June 1928, he requested that it be rescheduled. We do not know when this took place, but his approved dissertation was published on December 19, 1928 (*Tag der Promotion*).

The subject of his dissertation was *The Reductive Effects of Hypo-Phosphorous Acid on Organic Compounds*. The reductive mechanism of this acid (HPA for short) on inorganic compounds was well established and had a variety of industrial applications (Nickel plating and purification of nonferrous metals from their ores). Its effect on organic compounds had been studied since 1816, but there were no industrial applications developed. Targets were generation of intermediates for pharmaceutical products and purification of the polymer coloration process. The purpose of the research was to develop an understanding of the basic chemical mechanisms involved.

He was to investigate the possibility of esterification (sustaining a reaction of the acid with an alcohol, with ester and water as by-products); esters of HPA were considered to be promising intermediates for further chemical reactions. There had been no prior work on this topic. Neither was he able to make this process work, despite his knowledge of catalysis. At low temperature, the reaction would not proceed, and at high-temperature it was unstable. Today esterification of HPA is possible through several processes, but temperature control remains a critical parameter. Interestingly, these esters are building blocks of the backbone of DNA.

The remainder of his work was on the use of HPA to create effective reduction of several classes of organic compounds. He focused on five classes, in each on multiple compounds, and found that with the appropriate use of catalyst (palladium or copper), the great majority of reductive reactions could be sustained, such as reduction of nitro compounds to amines. Besides catalysts, pressure, temperature, agitation, and mercury buffer were other variables and techniques required to sustain the reactions and conduct the experiments. He established the chemical pathway along which these reactions proceed.

In a separate work, he alludes to the problem of mercury poisoning[8] and its adverse effect on memory, inflicting a professor (unclear whether this was Professor Thoms, Mannich, or another, but he investigated the issue by confirming that even at 37 °C, the body temperature, mercury adsorbs on Lecithin, a fatty compound rich in phosphorus and abundant in tissue comprising the nervous system). Mercury poisoning was reported in Haber's laboratory as well, which was also at Dahlem.[9]

Taghi Erani had several activities related to chemistry upon his return to Persia—he freelanced as an industrial chemist, developing a process to make body soap, marketed as *Darugar*. He also became a science teacher, chemistry and physics his main topics. He produced a textbook on modern chemistry (a copy is in the *Majles* library, 340 pages). When he was in government service, he used his expertise in chemistry in the Ministry of War. This is when he made his second return journey to Berlin in summer 1935 (forensic chemistry[10]), and later in the Ministry of Industry, in the assessment of mining operations.

Below are excerpts from articles he wrote on developments in radiochemistry, electrochemistry, and biochemistry:

Women are leading research in the subtlest field in science, the atom. Everyone knows Madame Curie, who passed away several months ago [July 1934]. Presently research by her daughter Irène Curie, in France, and lady-professor Meitner, in Germany, has created a significant step in our understanding of the structure of the atom....Irène Curie and her husband Joliot have discovered the property of induced radioactivity ... They bombarded the Boron atom with rays emanating from a compound of Radium...Critical observation was that if irradiation is removed, the Boron continues to emit rays, as if it has become radioactive...indeed it has turned into a radioactive isotope of a heavier element, Nitrogen.[11]

If the atomic nucleus dissociates, either naturally or induced by radiation, a tremendous amount of energy will be released, surpassing what we can produce in any power-plant. Perhaps one day use will be made of this energy of disintegration, which is an inexhaustible source of energy.[12]

The process which Nernst has analyzed in detail forms the basis of modern concepts in electrochemistry...As the temperature drops below a certain limit, the heat capacity of the substance drops (Dulong-Petit relation on constancy of molar heat capacity loses validity), therefore it becomes increasingly difficult to harness the electrochemical energy of the atom, and to proceed to lower its temperature. This makes it impossible to attain zero of absolute temperature. This observation is the foundation of modern thermodynamics (Heat Theorem).[13]

Until 1828, it was believed that it is not possible to synthesize organic matter. Then Wöhler obtained urea through synthesis. In less than one hundred years, organic chemistry reached a stage that Emil Fischer succeeded in making the most complex organic compounds, proteins, by linking eighteen amino acids obtained from decomposition of albuminoidal substances [1908]...At such rate, impossibilities of a practical nature fall into the realm of the possible. It is not far-fetched to envisage that one day science will manage to synthesize living cells.[14]

STRIFE IN PHYSICS, PHYSICS AND PHILOSOPHY

In 1926, Max Planck stepped down as the Professor of Theoretical Physics at the University of Berlin, where he had been teaching since 1889. He continued to give lectures at the university until 1928, and to the public and scientific gatherings thereafter, though this slowed down when the Nazis came to power in 1933 (there was an allegation that he is one-sixteenth Jewish, but it was dropped), his lectures came to a halt during the war (1939–1945), and he passed away shortly afterwards (1947).

In 1926, Planck's chair was offered to Erwin Schrödinger, one of the architects of quantum mechanics, a physical viewpoint that had begun with Planck and had reached a level of maturity to be able to describe both the wave and particle characteristics of matter, although a good deal still remained unexplained by this theoretical framework (nuclear and gravitational forces).

Planck had gained notoriety in 1900 when he postulated that emission of heat by blackbodies (bodies that absorb radiant energy but do not reflect it as light, rather give it off as heat by glowing) must occur not as a continuum but in discrete doses. Of course, the amount of heat given off depends on the amount of energy contained in the incoming radiation, but this heat was always a multiple of a fixed quantity, called quanta (a minute amount of energy). Planck maintained that this is the only conclusion we can draw from observations regarding radiation and heat. Precise experiments established the validity of Planck's postulate, with its quanta of heat or action, a theory which came to be called quantum physics. And in 1905, Einstein attributed the fact that a minimum intensity of light is required to generate electrical current when irradiating a metallic surface, to the quantized nature of light. The quanta of light came to be known as photons, and the associated phenomenon as the photoelectric effect. And in 1913, Bohr explained the fact that colors appear (red, aqua, blue, violet), when hydrogen gas filling a test tube is subjected to intense radiation (generated by an electric arc), to the fact that the radiation causes excitation of the electron that swirls around the hydrogen nucleus, and when this excited electron falls to a lower energy level, a fixed amount of energy is released as a packet of photons, corresponding to a specific color, each color corresponding to a narrow range of frequency or energy level, increasing from red to violet. Thus, he concluded that there must be four energy levels to which an excited electron in the hydrogen atom can drop, emitting a quantity of energy which is within the visible light range. Therefore, he attributed the discreteness of the spectral lines of the hydrogen atom to the discreteness of the energy levels or shells that can be occupied by the electron which surrounds the atomic nucleus. In other words, he attributed quantum properties to electron shells. Quantum physics was therefore established as a concise representation of the workings of matter, capable of describing the interaction of light with matter, photons with electrons.

When Taghi Erani arrived in Berlin in 1922, Planck was at the tail end of his academic career, though he embarked on a second, that of militating against revisionist concepts that had been creeping into physics, not by academic lightweights, but by certain prominent physicists (Mach, Jeans, Eddington). These scientists cast doubt on a whole array of established axioms of science—the existence of the external world independent of the observer; the existence of causative processes associated with natural phenomena; the certainty of being able to acquire verifiable knowledge of natural phenomena by careful observation and experimentation; the necessity of forming inferences beyond sense perceptions to design experiments to test our knowledge of natural phenomena and make it more precise. In short, they cast doubt on the self-corrective character of this process and the certainty of it resulting in a more complete understanding of nature, the possibility of making predictions as proof of the progressive nature of this learning process, and the prediction of behaviors of a statistical nature as proof of the law of causality (dynamic processes at the microscale as the cause of statistical regularity at the macroscale).

Planck's career in his youth had begun in adversity, and it seemed that it was destined to end in adversity in his twilight. When he had begun, there was little regard or recognition for theoretical physicists. Physicists, like chemists, were those who measured things or made things. Just as Ostwald had complained that many considered him not to be a chemist, as he had never synthesized a new compound; likewise, there was little regard for the theoretical physicist. Ostwald was one of the founders of physical chemistry, away from the limelight of organic chemistry, with its fixation on synthesis; so was Nernst, though Nernst had enough business sense to use his science to patent a light bulb concept (cerium oxide), which was superior to Edison's. He negotiated a million Gold-Mark royalty by licensing it to AEG. In a similar vein, Max von Laue, whose major contribution to physics (X-ray diffraction) came after Planck's theory of quants (1912 vs. 1900), received his Nobel Prize before Planck's (1914 vs. 1918), as he had the good sense to focus on experimental physics, though von Laue had been Planck's student and later was his assistant. Similarly, Einstein, when he received his Nobel Prize in 1921, it was not for the theory of relativity (1905, 1915) which had received observational confirmation during the solar eclipse of 1919 (light bent by the sun's gravitational field, double that predicted from Newton's gravitation), but for the more modest and tangible photoelectric effect.

Despite Planck's hardnosed approach to science and its philosophical foundations, there was a limit of how far he was willing to push his arguments, and this not for scientific considerations but for societal reasons. As someone whose life was scarred by Germany's bid to challenge its rivals and having witnessed how this had taken the country to the verge of destruction, he had come to the belief that religion, as the embodiment of faith and ethics, was the only savior of society, an antidote to its inherent instability and propensity to disintegration. Thus, while he defended the time-honored axioms of scientific investigation, he pronounced science and religion as not contradictory, but compatible and complementary. Planck stayed with his philosophy of the primacy of social cohesion by doing what he could when the Nazis came to power (including helping Jewish scientists and urging grassroots scientific work, as he did not believe in abandoning the ship).

Einstein took a different tack. He did express his philosophical views—famously as "God does not play dice" (in the debate on causality)—but not as actively as Planck, and he was not attached to Germany as such, seeing himself more as a citizen of the world, gravitating toward the non-assimilationist ideology emerging among certain European Jews and the pursuit of the Palestine project (differing from Rathenau and Haber on this issue, fellow Jews with whom he socialized).[15] He basked in the celebrity status that he had come to acquire, particularly after the May 1919 solar eclipse "bombshell." This was Einstein's approach during his Berlin years (1914–1932), before he immigrated to the USA and ended up lobbying for the atomic bomb.

Erani Commentary on Planck and Einstein, *Series in Exact Sciences*

In 1926, the year Planck took his retirement, Taghi Erani put to pen an ambitious plan to produce what may be called a synoptic survey of analytical science, as opposed to descriptive and taxonomic science (which according to him consisted of botany, zoology, geology, and paleontology). It was to consist of tomes encapsulating the principal theories and concepts in modern physics, chemistry, biology, psychology, and logic, emphasizing their interconnection as exemplified by boundary disciplines: physical chemistry, biochemistry, neurobiology, cultural psychology. He gave it the title *Series in Exact Sciences*, reminiscent of Planck's use of this term; description of this project came in 1500 words, below an excerpt:

> The reason to start the series with physics and pursue it with chemistry, biology, psychology, concluding with general principles or logic, is the nature of the evolution of thought. In any sequence, the more similar the initial and final states, the easier it is to comprehend the process. In mechanics, we deal not with the change of the object but with the change of its spatial and temporal coordinates. For this reason, it was not difficult for early civilizations to use the principles of mechanics for practical ends, such as the use of inclined planes by Egyptians to lift massive rocks to build pyramids. However, in transformative processes in chemistry, biology, and psychology, increasingly the similarity of initial and final states dissipates, giving to these phenomena the allure of mystery. But it is precisely in this sequence (of complexification) that our knowledge has grown, therefore we adopt this particular order for our series.[16]

It took ten years for this project to reach fruition, 1926–1935. The work began in Berlin, and it was completed after his return to Persia. (January 1926 came the plan for the series, as well as the first installment on physics, and December 1927 came the first installment on psychology, individual psychology, both works printed at *Kaviani Press* in Berlin; McGill University manuscript of *Psychology* has preface dated December 1927, Berlin.) His motivation for this project and the reaction it stirred in 1930s Persia will be described later. For now, we offer a sense of its scope.

Physics was covered in two volumes (Mechanics and Heat, 350 pages; Acoustics, Optics, Magnetism and Electricity, 550 pages), plus a treatise on *Theories of Science* or *Teorihaye Elm* (waves and particles, time and space, energy and matter, in thirty-eight topics, 74 pages, in effect Physics Volume Three); *Theory of Relativity* or *Farziye Nesbi* (drawing from Einstein's *Vier Vorlesungen über Relativitätstheorie* and Weyl's *Raum, Zeit, Materie*, 36 pages), and short articles on the *Four-Dimensional Atom* or *Atom va Boad'e Chaharom* and *Induced Radioactivity* or *Radioaktiv'e Masnoo'i* (total ten pages).

Chemistry was covered in one volume: inorganic, organic, and physical chemistry in condensed form (340 pages). Biology may be considered lost, although three articles on *Evolution-Adaptation-Inheritance* (*Takamol, Tabaiyat be Mohit, Ers*), Evolution of Living Organisms (*Takamol'e Mojoodat'e Zend'e*), and *Life and Mind* or *Zendegi va Rooh ham Maad'ist* are extant (total 33 pages). *Psychology* consists of one volume in two parts: Individual and Socio-historical Psychology (300 pages).

Logic, or more broadly philosophy and the history of ideas, is covered in various essays. *Man and Materialism* (*Bashar az Nazar'e Maadi*), examines man from the viewpoint of biology, psychology, and sociology (44 pages). *Mysticism and Materialism* (*Erfan va Osool'e Maadi*), surveys the rise of mystical and Gnostic ideas in China, India, Persia, Asia Minor, and Greece, and its manifestations in religion, philosophy, and science, the latter covering Europe, Henri Bergson particularly (50 pages). *Materialism and Dialectics* (*Materializm'e Dialektik*), considers the intertwined nature of matter-space-time and its manifestation in inanimate nature, living organisms, human society, and the mind, as a ceaseless process of non-directed change (50 pages). The essence of this is expressed in the ending of one of his essays:

> Nature and society is an interminable stream in continuous flux; tumult and turbulence alters its course from time to time; the new course in turn experiences upheavals and bank-blasts; only this play is endless and eternal.[17]

History of Science (*Tarikh'e Oloom*), covers Egyptian, Mesopotamian, Greek, Islamic, and modern periods (11 pages). *Historicizing Art* (*Tarikh-sazi dar Honar*) probes the interplay of societal evolution and art form (seven pages).

Therefore, the bulk of the series, which represents the bulk of his writings, was on the hard sciences, in which we find his commentary on the theories of modern science and its architects, some of whom he had occasion to witness firsthand by attending their lectures and exercise sessions (as listed in the *Lebenslauf* or vitae of his doctoral dissertation). He continued to monitor the Berlin scene, when he was back in Persia; below, samples of his commentary:

> The wave theory of Huygens and the theory of electromagnetics consider light as a continuum, while quantum theory expresses the properties of light in terms of discrete phenomena...But the two are being merged and the theory of wave mechanics is emerging from their synthesis, and proving the necessity of reconciling notions of continuum and non-continuum... Planck's work revolves around finding this common construct.[18]
>
> Planck, professor at the University of Berlin, believes idealism in physics, to which Mach and others subscribe, is not new...When Planck expressed this view in his discourse, there was bias in favor of Mach, and an independent thinker like Planck dared demonstrate the fallacy of predominant dogma. In his famous discourse, he states that reality exists independent of our thoughts.[19]
>
> While Planck subscribes to materialism in physical sciences and upholds the law of causality and the existence of nature independent of man...in the realm of social and psychological sciences he veers to the right and adopts a middle-ground between materialism and idealism. This stems from his sociological bias.[20]

And for Einstein, Taghi Erani had less of a soft spot:

> Despite the significance and scientific exactitude of the theory of relativity ...Einstein concluded from cosmological considerations that as curvature is proportional to density, the universe has to be self-enclosed and finite; once again we witness how sociological bias can influence a thinker...In recent years Einstein has abandoned this view, based on the reflections of de Sitter, the cosmologist. This stemmed from observation of distant stars that recede from us with incredible velocities, detected at Mt. Wilson [Coma cluster, expanding universe].[21]

His essays in which he explains the theory of relativity had conceptual and mathematical parts; in mathematics, he was attentive to modern geometry and its linkage to physics:

Non-Euclidean geometry is foundational to new physics, the theory of relativity and cosmology. Lobachevsky used hyperbolic [saddle] surfaces to postulate a notion contrary to Euclid's fifth postulate. Riemann's geometry is algebraic. ...Hilbert's geometry is axiomatic; these have a firmer foundation than Euclid's but are also less natural.[22]

For someone so enamored by science, it is difficult to imagine how he could have come to reach a stage that can only be described as heart-rending separation. Over time, science became a camouflage for the other passion with which he became afflicted; humanities, beginning with orientalism, passing over to psychology, and culminating in political philosophy and activism. And it was in Berlin that that passion was kindled.

NOTES

1. See Wilhelm II, *Kaiser's Memoires*, p. 85.
2. In 1915 Germans used chlorine gas in the battle of Ypres, the British did the same in the battle of Loos, and the French chemistry Nobel laureate Victor Grignard produced phosgene, which was used the same year by the French.
3. See Stoltzenberg, *Fritz Haber*, p. 165.
4. See Johnson, *The Kaiser's Chemists*, pp. 125–127 for an analysis of Fischer's speech.
5. Letter of Emil Fischer to Margarete Oppenheim, December 14, 1917, from *The Kaiser's Chemists: Science and Modernization in Imperial Germany* by Jeffrey Allan Johnson, p. 196. Copyright © 1990 by the University of North Carolina Press. Used by permission of the publisher.
6. Erani's academic history is given in the *Lebenslauf* of his doctoral dissertation.
7. See Heyns, *Chemistry at the University Level in West Germany*.
8. See Erani, *Donya*, 12-issue Volume, p. 285.
9. See Stoltzenberg, *Fritz Haber*, p. 68.
10. See Momeni, *Parvandeh*, p. 242.
11. See Erani, *Donya*, 12-issue Volume, p. 269.
12. Ibid., p. 238.
13. See Erani, *Osool'e Elm'e Chimi*, p. 101 (topic 46), p. 168 (topic 81).
14. See Erani, *Teorihaye Elm*, p. 319, and *Osool'e Elm'e Chimi*, p. 271 (topic 129).

15. For Rathenau's view, see Volkov, *Walther Rathanu*, p. 142. For Haber's view, see Stoltzenberg, *Fritz Haber*, pp. 296–297; for Einstein's view, see Seelig, *Albert Einstein*, p. 181 (1933), p. 189 (1938).

16. See Erani, *Osool'e Elm'e Physik*, Vols. 1 and 2 (both are prefaced by his 1500-word description of the *Series in Exact Sciences*).

17. See Erani, *Donya*, 12-issue Volume, p. 152.

18. Erani, *Teorihaye Elm*, pp. 320–321.

19. Ibid., p. 318.

20. Ibid., p. 321.

21. Erani, *Donya*, 12-issue Volume, p. 261.

22. See Erani, *Resalat fi Sharh ma Ashkal men Mosaderat Kitab Eqlidos*, passage in the text is the abstract of pages I-XIX (introduction).

CHAPTER 6

German Orientalists

Taghi Erani began dabbling in orientalism as soon as he arrived in Berlin in September 1922. His first contact with German orientalists, however, came in spring 1924, while his real exposure to German orientalism as a discipline and institution came in spring 1925. This would continue to the end of his stay in Berlin (December 1928). He worked in this domain on and off after Berlin, 1936 his final contribution, a year before his arrest and downfall.

That he began as soon as he hit the ground in Berlin was not some premeditated interest or passion for this field. Certainly, he enjoyed Persian literature; the only *Dar-ol-Fonun* professor he wrote about and praised is Mirza Abdol-Azim Khan Garkani, his professor of Persian literature.[1] He had also flirted with poetry in his youth, imbued with nationalistic emotions which was not hard to come by, given the dire condition of Persia. This was in contrast to its past, which mesmerized many, including Taghi Erani, though he shed that romantic perception of antiquity within few years of his arrival in Berlin.

His debut in orientalism was largely out of the necessity to secure a landing spot in a foreign environment in the midst of political turmoil, hyperinflation, flurry of refugees and war veterans. It was a lifeboat and mildly interesting, intellectually speaking. This motive drew him into the fold of the small but well-knit community of Persian expatriates in Berlin.

© The Author(s) 2019
Y. Jalali, *Taghi Erani, a Polymath in Interwar Berlin*,
https://doi.org/10.1007/978-3-319-97837-6_6

Best estimates are that there were two or three dozen Persians in Berlin at that time, a number that would rise to a hundred or so by the mid-1920s and remain at that level throughout the decade. The bulk of the Persians who were there had gravitated to Berlin during the Great War. Before the war, about thirty students had been sent by the government to various European countries, several to Germany. Other than that, we have little information as to who would have been there before the war (there was always the odd child of a merchant, one in Münster, 1907); we do know of course of the famed visitors who came to Berlin in the nineteenth century, both before and after German unification (1871).

Jean David Khan Massihi (surname meaning Christian) was the envoy who arrived in Prussia in 1850 to recruit instructors for *Dar-ol-Fonun*.[2] He was actually not Persian, but a citizen of the Ottoman Empire; however, he had spent his adult life working for the government of Persia. (He had acted as Amir Kabir's interpreter in early 1840s, during negotiations with the Ottomans over border disputes. This was the two-plus-two talks in Erzerum, overseen by Russia and Great Britain, which broke down as relations between the two superpowers degenerated, leading to the Crimean War.) Jean David Khan was of Armenian ethnicity and was deemed by Amir Kabir to create the least shock when meeting respectable fellows of the occidental world. He also knew the Persian scene quite well.

Several decades later, a Persian aristocrat, Mirza Malkom Khan, visited Berlin as envoy of Nasser-ol-Din Shah to participate in the Congress of Berlin (1878);[3] and the monarch himself passed through Berlin in all three of his state visits to Europe (1873, 1878, and 1889). These were attempts at triangulation, all of which failed, to engage the Germans through a loan or armaments deal to get respite from the Anglo-Russian straitjacket suffocating Persia. His last state visit created much interest in the German capital as an exotic ruler of the orient. Amusing cartoons appeared in the local press, showing him caught between a rock and a hard place, *zwei feuern*, two fires (Russia and England).[4] He received a stately reception from the Krupp corporation, photographs extant.[5] This trip, or rather its Vienna leg, became plot for a Joseph Roth novel years later (*String of Pearls*). Besides these anecdotal visits, there does not seem to have been any Persian community in Berlin prior to the Great War.

To get a sense of this small community, it suffices to glance at the profiles of a few of its leading figures; the rest seem to have acted as stage prop, leading normal lives. There was a store called *Persepolis* that sold accessories and memento.[6] Another called *Iranschähr* made a business out of theosophy and orientalism, by selling literature, offering seminars, and helping the occidental soul absolve itself of its sense of guilt and greed.[7] The principal of this outfit emigrated to Switzerland in the 1930s and became guru of a certain circle of enlightenment.

NOTABLE PERSIANS OF BERLIN

Of the notables, the most celebrated was Colonel Pasyan. He had passed through Berlin briefly during the war and had been deployed on the Eastern Front as the first pilot from Persia (1916–1918, before Persia had an Air Force), and had received the elite-level Iron Cross (worn on the left side of the chest). He was an ace pilot and had downed a number of enemy planes, was calm, cool, collected, and an otherworldly romantic figure. His favorite poet was not Saadi or Hafez, or even Rumi, but Rabindra Tagore:

> *If they answer not to thy call walk alone*
> *If they are afraid and cower mutely facing the wall*
> *O thou unlucky one*
> *Open thy mind and speak out alone*
> *If they turn away, and desert you when crossing the wilderness*
> *O thou unlucky one*
> *Trample the thorns under thy tread*
> *And along the blood-lined track travel alone*
> *If they do not hold up the light when the night is troubled with storm*
> *O thou unlucky one*
> *With the thunder flame of pain ignite thy own heart*
> *And let it burn alone*[8]

Indeed, these verses (poem called *Eka* or "Alone") turned out to be prophetic. He returned to Persia after the war (he stayed in Berlin over a year and left in early 1920), and he perished as the leader of a failed uprising in Khorasan province, after a rule of some six months, which gave a taste of social justice to the dispossessed (April–October 1921). He was from Tabriz, with ancestry from Caucasus, and his forename was a conjugated form of Taghi, and he was eulogized by Taghi Erani (ten years separated them in age), though Erani would have his own ideas of how to bring about the radical redressing of society that he felt was badly needed. Even General Ironside sounded the alarm at that time about the dire condition of Persia—the people bitterly poor, the elite rotten, the country ripe for communism.[9]

Pasyan had been in Persia's Gendarmerie, set up by the Swedes in 1910, force which sided with the Germans during the war, and took on the Persian Cossack Brigade, which sided with the Russians. Thus, we had Persians fighting Persians during the Great War. He showed enormous bravery during the war and gave a beating to the PCB, Reza Khan on the receiving side for he was assigned to the Hamedan *Otryad*, when Pasyan captured Hamedan in the battle of Musalla, the hill overseeing the city. When the forces of General Baratov poured into western Persia, he left for Constantinople and from there to Berlin. He participated in the German war effort as a private citizen and requested to be sent to the Eastern Front as he particularly despised Russia's molestation of Persia.

Another notable was Taghizadeh, not a romantic by any stretch of imagination, but a shrewd political operator. He was also from Tabriz. He had been a deputy in the first *Majles* when Persia became a constitutional monarchy (1906). When the *Majles* was bombarded (1908), he made it to London, rallied political support (with the help of Professor Edward Browne and the Balkan Committee of the House of Commons), and went back to Tabriz in the thick of the Sattar uprising, along with the English journalist Arthur Moore. But here he showed his true colors or political philosophy.

He suggested conciliation with the renegade king and got a beating from Sattar Khan's second in command in oblique Persian style. When news came that Russian troops were approaching Persia, he suggested to the *Anjuman* in Tabriz, the hub of nationalist resistance, that a telegram be sent to the king to plead for peace with the message: "the unkind father is preferable to the relentless foreigner." He was reprimanded by Bagher Khan, who said that it is curious that no one seems to be immune to gullibility, not even those of our own.[10]

After the fall of the absolutist king, he joined the second *Majles*, but political tensions and terrorism forced him to flee (1910). He made it to London once again and tried to lobby for the constitutionalist cause, but British foreign policy on this issue (Sir Edward Grey in charge) preferred *entente* with Russia rather than intervention on behalf of the nationalists. Indeed, in the dissolution of the second *Majles*, the ultimatum to expel the American financial administrators came not only from Russia but also from England. Having lost all hopes in the benevolence of the English, he went into his political wilderness.

This time, he surfaced in New York (1913), at Columbia University, doing legwork for Professor Abraham Valentine Williams Jackson, a renowned *Avesta* scholar. He also got a sense that the Americans were in no mood to take on the British and the Russians over Persia. Their Trojan horse had failed to accomplish its mission (William Morgan-Shuster and his team), and this fruit was not yet ripe to be picked (that moment would come four decades later, in 1953). Then, his fortunes turned.

He received a note from the German Consul in New York proposing a *tête-à-tête* (1914), behind which was Max von Oppenheim, self-appointed orientalist (archaeology his trade), Head of the Intelligence Bureau of the East (*Nachrichtenstelle für den Orient*). The Germans got wind of him through the Young Turks, as he always made his way to Europe via Constantinople.

In short time, he would be in Berlin (January 1915) and heading the Persia Committee to engage in propaganda on behalf of the German-Ottoman axis in the war, and not only by engaging in journalism (in Berlin, in *Shi'a* Iraq, and in Persia), but also by helping recruit militias in western Persia. And in fact, even more, to provide logistical support for the Niedermayer–Hentig expedition (attempt to connect Ottoman Iraq to Afghanistan through a wedge in the desert across Persia, in stealth from Russians in the north and Brits in the south, to entice the Emir of Afghanistan to strike Punjab, and to be received as liberator of India by other parties who were joining the scheme in India, dealing a fatal blow to the British).

He was kept in the dark about this expedition, but his operatives had to perform certain missions which were part of the larger blueprint for this initiative. He gathered around him a team of a dozen or so supporters (descending on Berlin from Constantinople, Paris, Cambridge, Lausanne), and the rest is history. The Germans were surprisingly courteous to support the Persia Committee until early 1922 (though from mid-1918 they were on shoestring budget), and then they had to pack up their bags and each go their own way.

Taghizadeh, as always, would land on his feet. Now, he threw in his lot with the rising strongman of Persia, Reza Khan, and would assist in brokering a trade deal with Russia, during 1922–1923 (concluded in 1927, Russia being Persia's largest trading partner, source of important customs proceeds for the government) and eventually became one of Reza Shah's loyal ministers. He was clever enough not to be too clever, as otherwise he would have been devoured by his master, who would eliminate anyone from his inner circle who could become his challenger, or challenger to his son, which was far easier (the crown prince being a teenager). Reza Shah had the misfortune of having a string of daughters before having a son. (He had eleven children from six marriages, all monogamous, only the fourth gave him a son, his fourth child.) He was nearly fifty when his son was born, and when he abdicated, his son was only twenty one.

Taghizadeh stayed in the game even after the fall of Reza Shah. He became a senator of senior standing and was there virtually forever. Taghi Erani had a particular dislike of Taghizadeh, and had met him in person (Taghizadeh had a German wife and dabbled in orientalism, Walter Bruno Henning his pen pal, and one of the books that Taghi Erani annotated[11] was brought from St. Petersburg to Berlin by Taghizadeh, when he was shuttling back and forth for the trade deal with Russia). Taghi Erani, in one of his political pamphlets (published anonymously), took sharp aim at the chameleon Taghizadeh, a constitutionalist turned right-wing royalist.[12]

Persian Expatriate Press in Berlin

As far as our story is concerned, what matters is that when the Persia Committee's print-shop was dismantled in early 1922 (it had been on *Leibnizstrasse 64*), its equipment was bought up at fire sale price by a Persian entrepreneur, another fellow from Tabriz (Abdul-Shakur Tabrizi), and he founded *Kaviani Press* at a nearby address, *Leibnizstrasse 43*. A few of the folks working with the committee also joined this new outfit, which was going to be run as a business, apolitical. Those who joined included Mohammad Qazvini who was a respected scholar of Persian literature, and Mahmud Ghanizadeh, another literary scholar (from Tabriz). One of the anecdotes is that when Qazvini, who was an erudite man, came back from a visit with the famed German orientalist Josef Markwart (at the latter's request), he was totally depressed. He said now I know that I know nothing. He is an ocean, me barely a drop![13] However, none of these scholars would stay around for too long in Berlin, after the Persia Committee was dismantled.

Two of the associates of Taghizadeh set up their own journals. One would eventually turn to theosophy as we alluded to earlier (Kazemzadeh, alias Iranschähr, the name of his journal, connoting land of Iran, he was also from Tabriz). The other was someone with literary ambitions, who became the father of short-story genre in Persian literature (Jamalzadeh, a native of Isfahan, witty with a biting humor, with a journal called *Elm va Honar*, Science and Art).

Taghi Erani worked with *Kaviani Press*, as well with these two gentlemen, for as long as their journals survived. He also collaborated with three other journals. Two were run by a fellow named Seif'e Azad (who had a pro-German agenda connected to a certain von Meyer in the security apparatus, journals called *Azadiye Shargh* or Liberation of the East, and *Sanaye'e Alman va Shargh*, or German Industry and the East). And there was another journal put out by the cultural society of students from Persia, called *Nameye Farangestan* or Letter from Europe, in which Taghi Erani was also active.

However, as Jamalzadeh reminisced years later (in 1973[14] and again in 1987,[15] the latter in two letters to the author), Taghi Erani fell out with most of the incoming students, who considered him as being too sure of his own ideas and intellectual prowess. Jamalzadeh, however, said that Taghi Erani was solid (*mohkam*) and knew his stuff, and academically he was spot on. Taghi Erani struck a friendship with a small group of the diaspora, with some of whom he would get engaged in political activities.

Jamalzadeh interacted with Taghi Erani throughout his stay in Berlin (initially due to their collaboration in journalism, and later because Jamalzadeh became the liaison officer in charge of student affairs for Persia's embassy; as Erani received state funding during the last two years of his studies, the two had regular exchanges). And in one of his trips to Persia, Jamalzadeh met Taghi Erani at his office at the Ministry of Industry in Tehran (autumn 1936), requested statistics about labor conditions in Persia (he worked for the International Labour Office in Geneva), as well as received copies of Erani's journal (*Donya*), and that was the last the two saw one another. Jamalzadeh, like his theosophist friend, left Germany for Switzerland in 1930s. He passed away in Geneva in 1997, at the age of one hundred and five.

GERMAN ORIENTALISTS

The year 1907 was another turning point in the Kaiser's quest to secure for Germany her rightful place in the elite club of the world's leading nations (continuation of a turn he took in 1890, when he sacked Bismarck). It is the year that colonialism, as practiced in Germany (since 1884), was to shed its prehistory of groping and guesswork, to become rigorous and scientific: a velvet glove cloaking the iron fist. In this élan, orientalism would be given free rein to prove its worth to the empire.

This new departure was not some intellectual revelation but came in the wake of a series of blunders on the scene of foreign and colonial policy such as the massacre of Herero and Namaqua in southwest Africa (present-day Namibia), which tarnished Germany's reputation as a peaceful nation.

But this change in course, like so much else in the Wilhemite era, was more a meander than a sharp turn. Thus, while a new breed of statesmen insightful about the world, particularly the orient, were brought in, the hawks continued to occupy important positions of power and responsibility. A co-habitation emerged.[16]

In 1907, Dernburg became the first colonial secretary; a banker and a liberal politician with an urbane touch in foreign affairs, a fig leaf to transition away from the genocidal affair in Namibia. He was succeeded by Solf, who was an accomplished orientalist (Indologist), an expert in Sanskrit, and the one who had created a model of so-called enlightened colonialism in German Samoa (1900–1910). Solf had spearheaded public works, social services, and job creation on the island. He was promoted to colonial secretary in 1911 and held that post until 1918, when the position was dissolved.

In 1914, when the war broke out, Solf pulled in Max von Oppenheim to create the Intelligence Bureau of the East (*Nachrichtenstelle für den Orient*), under the auspices of the Foreign Ministry, not only to gather intelligence but to foment anti-British insurgency in India, Persia, and forestall Allied attempts to carve out the Arab territories of the Ottomans, Germany's ally. Oppenheim was a freelance orientalist, an adventurer turned archaeologist and was intimately familiar with the Near East, notably Iraq, Egypt (where he had lived many years), and Syria (launched a major archaeological dig, near the Turkish border).

When Oppenheim was dispatched to Istanbul to buttress the Ottoman connection (1915), in came another orientalist, Schabinger von Schowingen, an expert in Turcology (Seljuq Persia his field of expertise, connoisseur of Nezam-ol-Molk, the grand Vezir). When he was dispatched to Jerusalem (1916), in came Eugen Mittwoch, Hebraist and master of Amharic, the Semitic (Ethiopic) language spoken in the Horn of Africa, adjacent to German colonies in east Africa. Mittwoch headed the intelligence bureau during 1916–1918 (after which it merged into other organizational forms), and he pulled in Kaufmann, yet another orientalist (though Swiss).

In 1918, Mittwoch was made director of the Oriental Languages Institute (*Seminar für Orientalische Sprachen*), which had been founded in 1887, at the tail end of Bismarck's chancellorship. The institute's first director had been Eduard Sachau, a renowned Semitist, and Mittwoch's doctoral advisor. The institute trained Foreign Service personnel, army officers, and entrepreneurs in oriental languages, with an enrollment peaking at thirteen hundred students, with a focus on the languages of the Near East (Turkish, Arabic, Persian), and languages of India, China, Africa, and eastern Europe.

This constellation of orientalists was connected to three other figures of importance to our story: Wilhelm Litten, Carl Heinrich Becker, and Friedrich Rosen.

Litten was a career diplomat and connoisseur of Persia, where he had lived fifteen years. He produced a number of books on Persia (all in German): Persia's New Constitution (1907), Persian Language Primer (1919), Persian Language for Diplomacy (1919), Colonialism in Persia from 1860 to Present (1920), Persia's Honeymoon (1925), and Persian Drama (1929), the latter prefaced by Friedrich Rosen.

During 1924–1926, Litten was instructor of Persian at the Oriental Languages Institute in Berlin, while between consular assignments. At the end of 1926, he received his posting as Consul in Baghdad, where he passed away a few years later (in 1932, he was born in 1880). Prior to Berlin, he had been Consul in Latvia (Libau: 1920–1924), Foreign Service staff in Bern (1917–1920), Consul in Tehran (1915–1917), Consul in Tabriz (1914–1915), Dragoman in Tabriz (1907–1914), Associate Dragoman in Tabriz (1902–1907), and trainee in Tehran and Constantinople (1902). Litten had lived in Tabriz while Taghi Erani had been there, including during the havoc arising in the wake of the constitutional revolution (Sattar's uprising against restoration of absolutism, 1908–1909), and the Great War contagion to Azerbaijan (October 1914) and the capital Tehran (November 1915). He had witnessed Persia's turn of the century convulsions.

Becker operated in a different universe, probing the nexus between orientalism and politics. He was a Hebraist like Mittwoch, but had made his name in Islamic Studies, having become a reputable Arabist. He founded the journal *der Islam* in 1910 (which runs to this day), and briefly occupied a chair in Middle East studies at the Colonial University in Hamburg. He had written his own job description for that position (meant to be purely philological, but he turned it into contemporary cultural studies), to which the university conceded, as he was respected in academia and had political clout.

Becker exercised his influence during the war from his base in academia. He supported the alliance with the Ottomans, but wanted them on a tight leash, the Armenian massacre lending credence to his judgement. And in the midst of the war, he joined Prussia's Ministry of Culture at a senior level. As testament to Becker's political savviness, he became Prussia's Minister of Culture (*Wissenschaft-Kunst-Volksbildung* or Science-Art-Education) after the war in 1921, and again during 1925–1930.

During the Weimar Republic (1919–1933), Prussia spanned the entirety of north Germany, which comprised sixty percent of its landmass (it was a Free State). As there was no equivalent ministerial post of culture or education at the federal level, this was a position that wielded great power. The Oriental Languages Institute was one of the organisms subordinate to Becker's ministry. This was an interesting twist, as during the war it had been Mittwoch (head of the Intelligence Bureau of the East) who censored Becker's critiques of the Ottomans (in Becker's words—"our censors are now more Turkish than the Turks"[17]). But now it was Mittwoch as head of the Oriental Languages Institute having to pay his respects to Herrn-Minister Becker.

FRIEDRICH ROSEN

Rosen was more a diplomat than an orientalist and had a versatile personality. He could connect with Litten, Becker, and Mittwoch genres of this world, as he could with larger-than-life characters like the Kaiser or Viceroy of India (he had tutored the children of Lord Dufferin), or a young aspiring scholar from Persia (Taghi Erani), engaging with each in their respective mental universe in total fluidity and naturalness. He had enormous range, a certain perspective on life, and his self-concocted principles.

Rosen had begun his professional life in orientalism, had been an instructor at the Oriental Languages Institute at the time of its founding, and had abandoned this role to pursue a career in diplomacy. Over a course of thirty-five years (1887–1921), this had taken him to Beirut, Tehran (where his son was born), Baghdad, Jerusalem (where the young Rosen had learned Hebrew, Turkish, and Arabic, when his father had been consul of the Prussian legation), Addis Ababa, Tangier, Bucharest, Lisbon, The Hague, and finally Berlin: the latter as Foreign Minister during May–October 1921.

Rosen, however, kept his fingers in orientalism all along, producing several works of note, such as *Persian Grammar*, which he produced as a collaboration of sorts with Nasser-ol-Din Shah (r. 1848–1896), before he was assassinated (Rosen was German Minister in Tehran during the end of the Nasseri period, using excerpts of the king's droll diary to illustrate contemporary colloquial Persian). In later years, he published the *Rubaiyat* or quatrains of Omar Khayyam in German verse and English prose (Rosen's mother was English, his father German) and published the newly surfaced quatrains attributed to Khayyam, in the original Persian (he had the financial wherewithal to publish a work bound to be commercially marginal, being in the original language). He also produced a German translation of Saadi's *Golistan* (Rose Garden, c. 1260 AD, blend of prose and poetry, earliest German translation from Adam Olearius, *Rosenthal*, 1654), as well as *Persia in Words and Image*, and an essay on the legend of *Harut and Marut* (sinful angles) which appears in the Old Testament and Koran (*Sura* of Cow).

Orientalism, therefore, ran in his blood. His uncle had been a notable scholar of Sanskrit and Arabic thriving in England (he died in his early thirties, translation of Khwarazmi's *Algebra* one of his notable works), and his father was a linguist, though not as scholarly as the uncle. His father signed up with the Foreign Service, beginning as Dragoman, progressing to higher grades.

In 1916, when Rosen was in Lisbon, Germany declared war on Portugal (German ships anchored there were immobilized, Portugal complying with British demands and joining the war on the side of the Allies), so he had to leave in short order. Back in Berlin, he was offered the ambassador position in Tehran, a post he rejected. This was not because he had occupied that post nearly thirty years earlier, but because he disagreed with the policy prescription that came with the job (oil production and adjacency to India made Persia strategic for both sides). Rosen was an advocate of "world politics without war." This was euphemism for *entente* with Great Britain. He had not been a Bismarck idolizer. In his memoires, he says that no serious-minded person could fail to see that change was needed, when Bismarck was sacked.[18] Nevertheless, there was a strain of Bismarck's political philosophy in him, prudence when it comes to Great Britain. Incidentally, this is the reason he was picked as Foreign Minister by Chancellor Joseph Wirth, because he thought Rosen could tame the relationship with the British, hence with the Allies, over the thorny issue of reparations. This proved to be a false assumption.

Rosen's main reason for refusing the Tehran job was because he believed the policy of fomenting insurgency in Persia was doomed to fail. And even if it succeeded, Germany would not be the beneficiary, as Persia's nationalists were using the Germans to defeat their colonial masters (Russia, Great Britain), not to switch masters. Eight years in Persia had taught him that. A different approach was necessary, something along Solf's doctrine of enlightened colonialism (Rosen and Solf had similar political outlooks, and in later years were part of the resistance to the Third Reich). However, in the midst of the war it was no time to bicker over these issues. His refusal was noted, and he ended up with the posting in The Hague.

Several years later—after the abdication of the Kaiser and his exit to Netherlands (where Rosen paid him a visit), the end of the Great War, the communist revolt in Germany and its collapse (execution of Karl Liebknecht and Rosa Luxemburg), and the beginnings of a wobbly Weimar Republic jolted by dozens of assassinations from extremist groups (right and left)—Rosen got the Foreign Ministry. When that position became untenable in a space of six months (May–October 1921), the London Ultimatum precipitating its end, he stepped down and threw himself into his orientalism. And it did not take long for his successor, Rathenau to be gunned down in bright daylight in Berlin (early morning of 24 June 1922), when Germany was drowning in hyperinflation, and French forces were about to cross the border into the Ruhr valley, the coal-producing province, Germany's gem.

Rosen became President of the German Oriental Society which put out the scholarly ZDMG Journal (*Zeitschrift der Deutschen Morgenländische Gesellschaft*), and he picked up his Khayyam. In 1922, came the fifth edition of his Epigrams of the Tentmaker Omar.[19]

This was Khayyam's poetry in German verse, which he had first published in 1909. And in 1930 came Khayyam's poetry in English prose.[20] A few months before the latter work came his memoires in English.[21] Between the latest German and English editions of Khayyam, 1922 and 1930, there appeared in 1925 Rosen's Persian edition of Khayyam,[22] specimens of *Rubaiyat* which had come to light, of all places, in Berlin. This is what served as the basis for the 1930 English translation. It was this work on the original Persian (analysis of provenance) that preoccupied Rosen for the better part of 1924, when he stumbled on a young man by the name of Taghi Erani at a Persian print-shop called *Kaviani Press* in Charlottenburg on *Leibnizstrasse 43*.

NOTES

1. See Erani, *Tadghighat'e Lesani*.
2. See Hashemian, *Tahavolat'e Farhangi'e Iran*, pp. 175–178.
3. See Raain, *Mirza Malkom Khan*.
4. See Hashemian, *Tahavolat'e Farhangi'e Iran*, p. 95.
5. Ibid., pp. 594–596 (note last numbered page is 576, pre-Appendix).
6. Persepolis, on *Goethestrasse 1*, Charlottenburg, was run by Reza Tarbiyat; see Behnam, *Berliniha*, p. 57.
7. Iranschähr, on *Martin-Lutherstrasse 5*, Berlin, was run by Hossein Kazemzadeh; his life account compiled by his nephew with the same surname; see Kazemzadeh, *Asar va Ahval*, p. 378.
8. Poem composed in Bengali (1905), English translation also by Tagore (1912).
9. See Ironside, *High Road to Command*, p. 153.
10. See Afshar, *Zendegi'e Toufani*, pp. 126–127, on Taghizadeh's account of Bagher Khan's reprimand.
11. This was *Vajhe-Din* of Nasser-Khosrow.
12. See condemnation of Taghizadeh in CRRP pamphlet, *Beyragh'e Enghelab*, in Chaqueri, *Taghi Erani dar Ayeneye Tarikh*, pp. 303–304.
13. See Afshar and Dehbashi, *Khaterat'e Jamalzadeh*, pp. 224–225.
14. See Golbon and Sharifi, *Mohakemeye Mohakemegaran*, pp. 201–203.
15. Private correspondence with S. M. A. Jamalzadeh.
16. Liberals consisted of—Bethmann-Hollweg (Chancellor, 1909–1917), Bernhard Dernburg (Colonial Secretary, 1907–1910), Wilhelm Solf (Colonial Secretary, 1911–1918), all advocates of détente with Great Britain. But power was in the hands of the "hawks"—Helmuth von Moltke (Chief of General Staff: 1904–1914), and Alfred von Tirpitz (Secretary of Navy: 1897–1916), both steered by Kaiser's vision for future of Germany as a naval and imperial power.

17. See Marchand, *German Orientalism*, p. 457.
18. See Rosen, *Oriental Memories of a German Diplomatist*, p. 84.
19. See Rosen, *Die Sinnsprüche Omars des Zeltmachers*.
20. See Rosen, *The Quatrains of Omar Khayyam*.
21. See Rosen, *Oriental Memories of a German Diplomatist*.
22. See Rosen, *Ruba'iyat-i Umar-i-Khayyam*.

CHAPTER 7

The Accidental Orientalist

Taghi Erani's debut in orientalism was as a proofreader (*mosaheh*) of Persian literary texts put out by *Kaviani Press*, which began operation in the second half of 1922, in westside Berlin, Charlottenburg, *Leibnizstrasse 43*. If *Kaviani* were similar to its predecessor, the print-shop of the Persia Committee operating during the war (*Leibnizstrasse 64*), it would have had a German typesetter, as there were no Persians with such skills in Berlin.

(There were many skilled in the art in Persia. The first trade unions in Persia were those of print setters, founded in 1906, and they were the most avant-garde, its members all able to read with ready access to wide variety of literature. They won the eight-hour workday for their members; amongst these, Erani would later find his most loyal followers.[1])

Typesetting was a tough task, and as anecdote would have it, the unfortunate technician familiar with Latin or Cyrillic orthography often went berserk over the morphology of the Persian script.[2] This would have been a difficult task even for a Persian typesetter. Besides the fact that many letters are subtle variations of each other, and most can take three forms, depending on whether they come at the start, middle, or end of the word, and there are no letters representing the principal vowels, only accents which may get dropped depending on context, there is the added dimension of penmanship.

The typesetting process began not with typewritten text but with handwritten script (Persian typewriters came after WWII), and even the tidiest handwriting produced knots, tough to untangle. Typeset samples

© The Author(s) 2019
Y. Jalali, *Taghi Erani, a Polymath in Interwar Berlin*,
https://doi.org/10.1007/978-3-319-97837-6_7

would be proofread time and again, but errors still crept into the final print. Thus, errata sheets were a standard feature of all publications, and these would have been Taghi Erani's first literary output.

Then in 1923 came two poems which showed he had a flair for versification, and in 1924 an article[3] which was a homage to his professor of literature at *Dar-ol-Fonun*, Mirza Abdol-Azim Khan Garkani. This article included an analysis of the linguistic shortcomings of the Persian language, in grammar and lexicon, and pragmatic steps that could be taken to address these gaps.

Here, he showed awareness of broader linguistic themes, the distinction between Semitic and Indo-European languages, and how Persian, though of the latter language family, has by force of history imbibed many elements of the former. This duality has engendered incongruities that prevent it from becoming a versatile language, a handicap, in view of the rapidly growing stock of knowledge emanating from modern European nations. The vessel of Persian language is too fragile to withstand the fluxes and flashes of modern knowledge.

In his essay, he also makes reference to the receding frontiers of the Persian language, accentuated since the late eighteenth century. Regions that were fountainheads of Persian literature (Samarkand, Bukhara, Ganja, on both aisles of the Caspian Sea) have experienced massive Russification, and the natives barely knew the language of their forefathers, much less the literary masterpieces that were conceived there.

Then, during 1924–1925 came his trilogy, which was a revival of rare works of prominent figures of Persia's literary renaissance (movement peaked from the tenth to the fifteenth century, after the Arab invasion of the seventh century), which showed not only the diversity but also the polarity in the thought universe of medieval Persia. Erani chose three literary figures from the eleventh, thirteenth, and fourteenth centuries and annotated one work from each to give a vista of the intellectual landscape of this period. It is in the midst of this project that he stumbled on Friedrich Rosen.

ENCOUNTER WITH ROSEN

Taghi Erani met Friedrich Rosen at *Kaviani Press* sometime in spring 1924.[4] What brought Rosen there was his Khayyam project. A body of Khayyam quatrains, over three hundred in number, surfaced at Prussia's state library in Berlin, and in addition, the Persian scholar Mohammad

Qazvini (formerly in Persia Committee, now collaborating with *Kaviani Press*) uncovered thirteen quatrains that were as old as any known specimens if not older (the scribe was a known poet of the fourteenth century[5]). This was an exciting project for Rosen, for it would make him the bearer of authentic Khayyam quatrains, not known to modern orientalists, and notably absent from the Bodleian collection at Oxford. Rosen had the scholarly knowledge, reputation and financial wherewithal to pull off the project. All projects required an initial outlay of capital, and returns were never guaranteed, especially for a work in the original language. Buyers were a small circle of enthusiasts—orientalists, bibliographers, and literary diehards of the diaspora.

However, compared to Rosen's previous work on Khayyam, which was a selection of known quatrains rendered into German verse, the new undertaking would be a major feat. The literary scholar, Mohammad Qazvini, and the enthusiastic youth, Taghi Erani, lent their support to this initiative.

During the flurry of this activity, Rosen used his influence to have a rare work of Khayyam sent from libraries in Leiden (the Netherlands) and Gotha (Germany) to Berlin, to be transcribed for later annotation and publication.[6] This was Khayyam's critique of Euclidean geometry (c. 1070 AD). Rosen was of the belief that one obstacle to the mystery of the *Rubaiyat*—were these echoes of Epicurean materialism extolling carnal pleasures or were they mere allegories of mysticism—was insufficient knowledge of the life of its author. Rosen believed that this could be remedied in part by scrutinizing Khayyam's known works, specimens of which were slumbering in medieval libraries of Europe.

Rosen possibly had another motive as well. He had great admiration for his late uncle whom he had never met, as he had passed away at the onset of his prime, twenty years before Rosen was born (he was thirty one when he died, 1805–1837), and had produced a rare work of Islamic mathematics, rendering it from Arabic into English in 1831 (though the mathematician was Persian and the translator a German specializing in Indology).[7] Dismayed by the way his political career had ended (though he did make Foreign Minister, despite his Jewish background, a handicap in officialdom even for those who converted to Christianity), Rosen was on a mission to make what imprint he could in the realm of orientalism. And as the medieval orient had been the guardian of science of antiquity, and a source of new knowledge in its own right, a serious theme in this realm would be highly regarded by *connoisseurs* of the art. It would also be a nod to his late uncle.

Rosen recognized that such work required not only knowledge of orientalism but also knowledge of science; and in this case a science that had come into the limelight of man's modern conception of the universe—modern geometry, an offshoot of ancient geometry, providing the bedrock of modern physics and cosmology. As is well known, even Einstein did not completely grasp the mathematical formalism attached to his general theory of relativity, the basis of modern cosmology. Of the work of Hermann Minkowski, his former professor, he famously said—I do not understand it myself.[8]

Taghi Erani was perhaps the closest Rosen had to his image of an ideal partner for such a project—familiar with medieval Persia and Khayyam and well versed in Arabic. Taghi Erani had received a linguistic grounding in Arabic from early childhood at the initiative of his mother via Koranic classes. Khayyam's work, as all major works of scholarship and administration of that period in Persia, was in Arabic. The renaissance of the Persian language made inroads only in literature. Erani was also a serious student of fundamental science, his professors the pathbreakers of modern science: Nernst, Bodenstein, Freundlich, Planck, Einstein, Pringsheim (Peter), the last three most relevant to modern physics and cosmology.

This, in addition to the fact that the young man had good command of five of the seven languages mastered by Rosen—German, French, Persian, Turkish, and Arabic—with a rudimentary knowledge of a sixth, English, reflected in correspondence and sources cited in his works, must have clinched the deal in favor of Taghi Erani to be proposed this exceptional collaboration opportunity. (Rosen also knew Hebrew.) Already in 1925, Taghi Erani would be busy at the Prussian state library, transcribing the work by photographs and handwritten copying, the results safeguarded at the library, until such time as he could make a careful study of the material and prepare it for publication in consultation with Rosen.

Further, as archives reveal, Rosen opened doors for the young scholar to land a unique teaching position at the Oriental Languages Institute (*Seminar für Orientalische Sprachen*). That is, not to teach Persian, but to teach comparative rhetoric, comprising Persian, Turkish, Arabic, delivered in German, making for a wider audience: a domain reserved for scholars. Mittwoch, head of the institute, referred to Taghi Erani as *"persische Gelehrte"* (a Persian scholar) in his correspondence with Carl Heinrich Becker, Prussia's Minister of Culture.[9]

THE THREE MUSKETEERS OF MEDIEVAL PERSIA

A Persian variant of *The Three Musketeers*: a theologian, a jester, and a sage of medieval Persia, were characters that Taghi Erani got to know intimately because of his publication projects at *Kaviani Press*. This was his debut into the humanities, indeed into the thorny subject of human nature. Three works, one from each one of these literary figures, were published, the first two before Erani began teaching at the Oriental Languages Institute, and the third shortly afterwards, as alluded to in its preface, signed *"T. Erani, Instructor of Orientalism at the University of Berlin, 23-September-1925."*[10]

From one vantage point, these were works from a bygone era, notably there was a work considered by Professor Carl Ethé to have been lost to posterity (described in *Grundriss der Iranischen Philologie*); thus, the manuscript had to be analyzed for its linguistic characteristics, correspondence with extant works of the presumed author, and descriptions contained in younger works proximate to the *oeuvre*, by medieval bibliographers. From another perspective, these works contained stories that could not not arouse the curiosity of the reader, particularly Taghi Erani, who had begun to take an interest in humanities as a diversion from physical sciences.

One of the musketeers was ideological in his attitude to life. If the right ideology could be spotted, and if one could infuse the mind of men by such ideology, a radically different reality could emerge. There was a bit of Don Quixote in this character, though he belonged to a different universe than the creature of Cervantes—not chivalry but piety, not romantic heroism but theological humanism, not wielding the lance but stroking the pen. One thing they did share, however, was travel to distant lands, causing them to witness the intriguing tapestry of life. This character was a theologian–philosopher with a tenure in civil administration, not foreign to realities of life, yet predisposed to idealism, motivated by the belief in the nobleness of man.

The second was the polar opposite of the first, propounding that humans by nature are depraved. They are self-serving creatures to which culture and education gives only the veneer of civility. If they master one thing, it is the art of pretense. If a man is penniless, it is not because he is honest—though it is true that all men of means are dishonest—but because he has failed to be dishonest by his ineptitude. Thus, he goes a step beyond Machiavelli who lamented to a patron in Rome—"my poverty bears witness to my fidelity and honesty."[11] For this musketeer, who is a jester in a medieval court of second-rate importance in Persia, the only thing poverty bears witness to is incompetence in the art of swindle, swindling the hearts and minds of men into gestures of patronage and generosity. That he is a jester is not because he is a lightweight in a cerebral sense. The only reason he ended up so is because he lost out in court intrigue to rivals who were more cunning, though inferior in the art of poetry and prose, not to mention intellect and powers of perception.

The third character is situated between these two, both in philosophical temperament and historical coordinates. The first died a fugitive away from his home around 1090 AD. The second expired around 1370 AD, on his deathbed playing a last prank on his children. The hidden treasure to which he pointed in waning whispers contained a note—*truth be told, your father did die a pauper; may you prosper in life through your own industry!*[12]

The third character passed away around 1290 AD and has the merit of being considered a sage in the Persian culture, not because many an illiterate person know his aphorisms by heart, but because even those who have trouble reconstructing the rhyme and rhythm of his words act largely in ways that he painted, in accordance with human nature, as he perceived it.

Having been a vagabond for the better part of his life, and having lived a long life, by some accounts exceeding one hundred, and having turned his observations and intuitions into writing in the latter part of his life, what he wrote was not academic knowledge, rather hard-earned lessons of life, elaboration of which could be made with countless anecdotes.

He had left his hometown of Shiraz for Baghdad, having lost his father at a young age, and became the pupil and later a tutor at *Nezamiye* college. He travelled widely while based in Baghdad, but fled before the arrival of the Mongols, led by Hulagu Khan (grandson of Genghis Khan) who sacked Baghdad (1258 AD). He had seen the rule of the Crusaders in Jerusalem (Second Kingdom, 1192–1291 AD). He was enslaved by the Franks in chain gangs with Jews in Tripoli (Lebanon) and was bought by a man of means from Aleppo to be wedded to his demonic daughter (from whom he fled). He was chased in temples in Kashgar by Buddhist monks thirsting for his blood. He roamed through

Persia, Arabia, North Africa, Malta, Levant, Caucasus, Khorasan, and Transoxiana. Finally, he returned to his birthplace of Shiraz, not far from the ruins of the Persian Empire, its collapsed columns and mustiness humbling food for thought.

For him, it was not unseemly to lie if it meant the lie stemmed strife (*well-meaning lie, better that strife-stirring truth*[13]), and it was fine to kill as well, if the alternative was self-demise. In Kashgar, he had hurled a monk into a well and had fled. The sage had discovered the trickery of the monks to make believe the animation of their giant idol. Earlier he had questioned the wisdom of worshipping a giant block of wood and now had discovered what was not to be revealed, hidden ropes and pulleys attached to the idol. He was spotted by the guardian who ran off to alert the congregation; so, he chased the guardian, pushed him into a well, the passing *caravans* on the routes of Turkestan his getaway.[14]

In the sage's school of thought, although experience suggests many things about human nature, especially propensity to backbiting (*none learn of me the science of the bow, who in the end make me not their target*[15]), the grandest truth is this—as varied as the clans of the human race are, all are limbs of the same trunk, when one branch is stricken with disease, it is the whole tree that is smitten for its life. Be tolerant of others, no matter how different they seem to be from you; beneath, all are the same, equally capable of vice and virtue.

Thus, his prescription:

- To the kings and rulers: be just and generous; just as the pawn can become a Vezir on the chessboard, so can a slave become a king, and is there not an army of slaves among your ranks? Poverty borders on the denial of god, denial pathway to dissent, dissent the end of kingdom; and is it not true that it is by pruning that yields more the vine? What wisdom in hoarding?[16]
- To the clergy and man of religion: pretend less, fear more the Almighty. Religion is not the rosary and the prayer rug, wear the hat of the tartar, keep the heart of the humble.[17]
- To the common man: be content and pragmatic. A Christian's well may not be pure, it is true, but it will do to wash the carcass of a Jew.[18]

Such are the laws of conduct in life, according to the sage of medieval Persia.

***

The sage was none other than Saadi of Shiraz—known for *Golistan* (The Rose Garden), *Bustan* (The Orchard), *Qazals* (Odes), *Qasideh* (Elegy), and other works. Taghi Erani's publication was a selection of Saadi's Odes (called *Badaye*, or Startling Truths, published September 23, 1925),[19] which was authenticated by Professor Lucas White King of Belfast, a former administrator of the British Crown in India, who passed away toward the end of this project. A short biography and eulogy is given by Taghi Erani, who refers to Professor Edward Browne's *Literary History of Persia* to remind the reader of the stature of Saadi in the genre of Odes—"*In his Qazals, or Odes, as already said, Saadi is considered as inferior to no Persian poet, not even Hafez.*"[20] Elsewhere, he thanks Professor Browne for financial assistance to such literary projects.

The jester was Obeyd of Zakan, who also flourished in Shiraz, though originally from Qazvin (hamlet of Zakan). In Shiraz, he was a contemporary of a younger poet of great talent named Hafez (who inspired Goethe's *Divan*) and with him was caught in a love triangle with the daughter of the local ruler, the poetess Jahan. The crafty rhymester who made life difficult for Obeyd was Khaju of Kerman, poet laureate. It was a collection of Obeyd's satirical works, *Lataef*, or soothing tales, that was edited and prefaced by Taghi Erani, published by *Kaviani Press* (January 5, 1925).[21]

And the theologian was Nasser-Khosrow, a civil servant in the court of Seljuq rulers of Khorasan, before they overtook Persia and part of present-day Turkey.[22] His midlife crisis led him to abandon the secure but uneventful life of a provincial civil servant and embark on a journey that would last six years (*Safar-Nameh*, his best-known work, a travelogue). Its high point was the revelation of the sagacity of the Fatimid rule in Egypt and North Africa.

Thus, he journeyed back to his home province of Khorasan to become a missionary of the Fatimid faith, through the art of dissimulation (Seljuqs were in alliance with the Caliph in Baghdad and of *Sunni* faith, while Fatimid of Egypt were of Ismaili faith, a branch of *Shi'a* Islam, and were political rivals of Baghdad). In this endeavor, he did not have much luck, for at this time another offshoot of Ismaili faith was resorting to high-profile assassinations launched from the safety of their strongholds, to hold their own against the Seljuq rule. They had a chain of forts which were run by an order with iron discipline, led by Hassan Sabbah. The Ismaili followers were persecuted by the Seljuq, including Nasser-Khosrow, who became a fugitive and died far from his home province.

He was born in 1004, his journey took place during 1046–1052; he became a fugitive in 1063 and died in 1088.

The work produced by Taghi Erani was called *Vajhe Din* (Way of Faith)[23] and had been found among Ismaili communities in Bukhara, and from there had been expedited to the Asian Museum in Petrograd (St. Petersburg). In Petrograd, the work was photographed by Taghizadeh, former head of the now-defunct Persia Committee, on visit to Russia to broker a trade deal on behalf of the new government in Persia (under Reza Khan, as Head of Army, Minister of War, and Prime Minister), who brought it to Berlin. There, it was examined by two scholars of Persian language (Qazvini and Ghanizadeh), and the work was compiled, edited, and prefaced by Taghi Erani, published by *Kaviani Press*. Analysis and commentary was provided as to why the work is considered to be authentic, as Professor Ethé had considered this work by Nasser-Khosrow to have fallen prey to the ravages of time. One regret expressed by Taghi Erani was that they had no recourse to compare this manuscript to a second, also kept in Petrograd.

The chief idea of this work was the concept of temporal representation or *imamate*, and how this devolves to enlightened rulers of the time, by implication to the reign of Al-Mustansir Billah in Egypt, who ruled for sixty years. This is not explicitly stated, and the book is cloaked in much theology of creation, the philosophy of time (considered beginning-less, creation being a potentiality before becoming a materiality), numerology (number seven being significant), and religious rituals (down to personal hygiene). As a result, Nasser-Khosrow used methods of occult creeds, signs and symbolism, to convey his meaning. Only after he was discovered by Seljuq informers did he become a fugitive, his spirit apparently never broken (intransigent poetry). He has about a dozen works and a *divan* of poetry.

LECTURESHIP AT THE ORIENTAL LANGUAGES INSTITUTE, COMPARATIVE RHETORIC

The Weimar archives reveal how Taghi Erani landed the teaching job at the Oriental Languages Institute while he was in the midst of his trilogy:

January 14, 1925, 1300 hours—Wilhelm Litten receives a phone call from Taghi Erani, triggering a sequence of events, instigated by Friedrich Rosen. According to Litten, Erani demonstrates striking mastery of

German, with nearly flawless speech (*"eine auffallende Beherrschung der deutschen Sprache zeigts, die er fast fehlerlos sprach"*).[24]

January 14, 1925, 1500 hours—Taghi Erani arrives at *Dorotheenstrasse 7*, the Oriental Languages Institute, for a personal interview with Wilhelm Litten.[25]

January 16, 1925—A three-page memo from Consul Litten to Direktor Mittwoch describing details of the interview: Taghi Erani is a student from Persia, studying natural sciences at the University of Berlin, majoring in chemistry. He speaks a pure Persian with an accent from Tehran. His father is a civil servant working on the board of the tobacco tax department. He proposes two series of lectures to be delivered in German for advanced students. This does not compete with the courses on offer; rather, it is a good complement. First is a series of lectures on oriental rhetoric, style, and logic derived from his knowledge of Persian, Arabic, Turkish literature. Second is a series of lectures on Persian literature based on samples from prominent figures with unique styles. Given that the lectures will be delivered in German, it can interest a wider circle than those studying these languages. He has left a fine impression on me (*"Schon heute kann ich aber sagen, daß Herr Taghi Erani auch auf mich einen guten Eindruck gemacht hat"*).[26] I cannot judge the viability of financing.

January 27, 1925—Letter from Taghi Erani to Prussian Culture Ministry asking for permission to offer a course in oriental rhetoric, style, and logic at the Oriental Languages Institute. He attaches a bio and indicates that additional information about him can be obtained from His Excellency Rosen (*"Weitere Auskünfte über mich kann S. Exz. Herr Dr. Rosen erteilen"*). Signed T. Erani.[27]

January 30, 1925—Letter from Direktor Mittwoch to Herrn Minister Becker. Reichminister Dr. Rosen asked me for a special task concerning a student from Persia by the name of Taghi Erani. At my instigation Consul Litten evaluated him, and I also met him, and he left a fine impression. I provide a description of his proposed course and linguistic abilities. If approved, he would teach two hours per week. I recommend that this be granted (*"Ich darf daher angelegentlichst empfehlen, Herrn Taghi Erani mit der Abhaltung der genannten Vorlesungen in zwei Wochenstunden betrauen zu wollen"*).[28]

April 22, 1925—Handwritten letter from Minister Becker to Direktor Mittwoch regarding the application of Taghi Erani: proposal approved to offer this course and appoint Taghi Erani for the job. (Lapse of time from Mittwoch's letter of end January, most likely due to background check, causing the forthcoming academic term to be nearly missed.)[29]

October 20, 1925—Letter from Direktor Mittwoch to Herrn Minister Becker, in reference to correspondence of April 22, 1925: I am honored to report that the scholar from Persia (*"persische Gelehrte"*) Taghi Erani has begun since May to teach a course in oriental rhetoric, style, and logic, two hours per week (Monday and Thursday, 7–8 p.m.). This was not included in the summer semester directory for lack of time. Five hours of lectures has been observed. I can confirm the lectures are proceeding successfully. As compensation, he is to receive 47.25 marks per month. He has thus far been paid for three months (141.75 M). I explain the merit of this course and its synergy with our language courses in Persian, Arabic, Turkish, where students become familiar with the thought processes in these languages. Requesting that funding be granted for its continuation.[30]

November 19, 1925—Handwritten note from Minister Becker to Direktor Mittwoch approving the request for continuation of the seminar series by Taghi Erani. (It ended up being delivered eight semesters from summer 1925 to winter 1928; eighth, however, was interrupted in December 1928, instead of the scheduled ending of March 1929, due to his sudden departure for Persia.[31])

***

We do not have precise detail on what Taghi Erani covered in his course. We have Litten's description of his proposal which would cover logic, grammar, and rhetoric. It appears that his approach was to take passages from works of well-known literary figures and articulate these themes within its specific context. It is possible these were done somewhat interactively, as he does make a point about using modern methods of pedagogy in his teaching.

Neither do we know the extent to which Rosen influenced the content or format of presenting this material. What we do know is that among war-fatigued orientalists, there was a desire to return to pure orientalism, as opposed to applied (politically exploitable) orientalism, so-called furious orientalism, to use Professor Marchand's characterization. Thus, articulating elements of semantics and style, lexicology and phonology, rhetoric and aesthetics, would have resonated with the mood and mind-set of postwar times.

As we plough through his works, we observe that he expanded his repertoire beyond his initial work on the literary figures of medieval Persia. He examined *Grundriss* encyclopedists (Bartholomae, Geiger, Ethé), Markwart, Christensen, Nöldeke, medieval historians (Tabari, Bayhaqi), to gain a clearer understanding of the development of thought in Persia from Zoroastrian times to Islam, to the more recent era. He also expanded his geographical scope, developing familiarity with the intellectual history of China (warring states period, Mohism, Taoism, Confucianism), India (Lokayata, Jainism, Hinduism, Buddhism), monotheistic religions, Greek philosophy, and Western intellectual development (scholasticism, enlightenment, romanticism, materialism). This led to the history of science and to modern science, which merged with his project in *Series in Exact Sciences*. Parts of this exploration may have entered into his lectures at the institute.

DEPARTURE FROM THE INSTITUTE, THE ENDING WITH ROSEN

As reflected in the archival correspondence, Taghi Erani began teaching at the institute in May 1925, in the summer semester, which officially ran from April 16 to August 15. The approval for this seminar and appointment of Taghi Erani came from the Ministry of Culture on April 22, so he had a late start. During this first semester, his teaching was observed, as alluded to in Mittwoch correspondence, most likely by Litten, who had excellent knowledge of Persian literature (he wrote *Das Drama in Persien* in 1929, prefaced by Rosen), as well as knowledge of Arabic (he became Consul in Iraq, 1927–1932). Therefore, we may surmise the two had a cordial professional relationship.

Litten stayed on for another two semesters (winter 1925 and summer 1926), and right before the start of the winter semester, 1926, he received his letter of assignment for his posting in Baghdad and left immediately. (Litten had been on loan from the Foreign Ministry to the Ministry of Culture, while between consular assignments.) Thus, suddenly there was no instructor of Persian at the institute. The archives contain Litten's note to Mittwoch (October 15, 1926), stating that he is to leave immediately, regretting that he would not be able to teach the course, due to commence the following day.[32]

Then came a flurry of memos, one by Taghi Erani (October 28, 1926),[33] addressed to Minister Becker, stating that he is an instructor at the institute, teaching his subject with modern pedagogical methodology (for which he has been commended by Consul Litten), and requesting to be considered for the post of teaching the Persian language. He also states that His Excellency, Rosen, considers he would be the most suitable candidate for this position.

It then took four months for a final decision to be made on this issue (February 23, 1927), all along while there was no teaching of Persian at the institute. The decision was made not by the Ministry of Culture but by the Foreign Ministry, proposing one of their employees, a certain Abdullah Sebastian Beck to take this position. It was explained that Mr. Beck, who is an Afghani national, had worked in the Intelligence Bureau of the East from 1915–1921 (*"Von 1915 bis 1921 ist er in der Nachrichtenstelle für den Orient tätig gewesen..."*[34]). This end date is intriguing as officially this service was dissolved in 1918, after the war, but evidently it must have continued operation, subsumed in other organizational forms. In the directive, it is explicitly stated that the candidacy of Mr. Taghi Erani for this post is not supported by us. (*"Den Antrag von Herrn Taghi Erani bittet das Auswärtige Amt nicht zu unterstützen."*[35])

Whether or not Taghi Erani knew these details, we do not know. But certainly, he must have sensed that in such matters other considerations were at work; that there was a stark contrast between the aims of high scholarship and academic orientalism, on the one hand, and its practical exploitation in the political realm, on the other.

It was perhaps for this reason that in later years his attitude toward this field was highly guarded. In the orientalist studies of Persia, he only considered select works to be of scientific worth, such as those of the *Grundriss* encyclopedists (Bartholomae, Geiger, Ethé), Markwart, Nöldeke, and Christensen.

The archives also reveal the suddenness of his departure from Berlin. From sources related to his political activism, we know more of the story; but in terms of orientalist archives what we see is his petition (dated December 10, 1928)[36] to Minister Becker requesting a six-week leave of absence for an urgent trip to Persia, and memos leading to the approval of this request (January 23, 1929), by which time he had already left Berlin (end-December 1928). Archives also describe the information Mittwoch obtained from Persia's embassy in Berlin about the departure of Taghi Erani, to the effect that he was recalled by the government due to his political activities and that he would not be returning to Berlin. Then, an exchange of notes between Mittwoch and Becker led to the decision to end his association with the institute (April 13, 1929).

Taghi Erani's departure was so sudden that he did not even take with him the transcribed copies of Khayyam's critique of Euclid, kept at the Prussian state library. Perhaps he thought he would be returning shortly. He was able to recover these during his first return trip to Berlin in autumn 1934. It is uncertain whether he met with Rosen during this trip, as Rosen may have already left for Beijing, to spend time with his son who worked at the German legation in Tsing-Tao, and to be away from the rising tide of anti-Semitism at home, after Hitler's rise to power (January 1933).

The Khayyam work on the critique of Euclid was published by Taghi Erani in Tehran,[37] with a twenty-page introduction in Persian and a shorter one in Arabic, positioning Khayyam's critique of Euclid within earlier and later critiques (conforming to the description given by Hermann Weyl[38]), culminating in the emergence of non-Euclidean geometry in the nineteenth century (Lobachevsky, Riemann, Hilbert), which forms the mathematical foundation of modern physics and cosmology. This came out on February 1, 1936 (*Imprimerie Sirousse*, Tehran); but Rosen had passed away two months earlier in Beijing (November 27, 1935). Taghi Erani, who was evidently in touch with Rosen, wrote in his preface that until about a month ago, this old friend was eagerly awaiting the publication of this work; and he waited for ten years. Alas as this work dawns, he has eclipsed. He gave a moving eulogy of Rosen and a synopsis of his contributions to orientalism.

Taghi Erani's View of Orientalism, Shift Toward Psychology

There are several themes concerning orientalism that surface in the writings of Taghi Erani. One is that he seems to view orientalism as a branch of historical research, belonging to cultural history, which requires not only an understanding of philology, but also an understanding of the material culture of the period in question, its historical context, and intellectual *zeitgeist*:

This sort of inquiry cannot be done merely by knowing languages [philology], because the intent is not translation, rather scientific investigation...A manuscript has to be examined from the perspective of chronology, phraseology, vocabulary, and subject-matter. Conditions of life where it originated, its antecedents...the influence of local environment, material conditions and temporal characteristics on the work and the influence of the work on the environment and its milieu must be examined.[39]

He takes issue with approaches that attribute the work of a literary figure to exceptionalism, timeless qualities and genius, treating the subject matter divorced of its historical context and interconnections:

Utterings such as 'real genius is timeless,' 'essence of art is beyond social imprint,' are not unheard of in corridors of faculty of philosophy, theology and literature. One of this ilk has written a book in German called 'Art Beyond Time'...Another, a certain Gundolf, has filled a whole book about Goethe with similar propositions, despite the fact that Goethe believed in the impact of social setting on thought formation of the thinker....The same forces that brought about the revolutions of 1848 created the literature that anticipated it in the 1750s.[40]

Another attribute is his critique of icons of Persia's literary history, averring their irrelevance to contemporary life. There is a strong tendency toward iconoclasm. This includes pantheonic figures whose works he had treated with approbation in his youth:

Nasser-Khosrow (eleventh century) claims the materialist philosophers of Persia (Iranshahri, Razi, Bahmanyar, eighth and ninth century), have gone astray by arguing that matter (*Hayuli*) is non-created and is eternally ancient (*Qadim*). He says such claims are not only contrary to the word of God, but their uttering is tantamount to proclaiming prophethood. Nasser-Khosrow then argues that as the assertion of the one whom everyone attests as the true prophet [Mohammad] is in contradiction to the assertion of the materialists, then the claim of the materialists is false. Works such as *Vajhe Din* and *Zad-ol-Mosaferin* are replete with such ludicrous logic.[41]

Let us see what conditions during the time of Saadi (thirteenth century) resulted in a work like *Golistan*. Hordes of Mongol Il-khans gained mastery over society. These had no skills other than the art of killing. They idled while the people were condemned to toil. Then Saadi emerged, with those words of eloquence, heart-throbbing prose and poetry, discourse on virtues of a life of contentment, ways of the *dervish*, and manners befitting the mighty. On the one hand to counsel the privileged towards righteousness and on the other to lighten the emotional burden of the toiler. For his era, Saadi fulfilled his mission superbly. But in the twentieth century, this manner of thought is ill-fated...The poems of Saadi and Molavi [Rumi] and Hafez, those works of the bygones that teach the art of contentment and introversion, today hurl the youth towards resignation, drugs, alcohol, and decay of morals.[42]

If you glance through the *Divan* of Hafez (fourteenth century), you see in every page at least five to ten times words like wine, wine-bringer, tavern, spirits, goblet and so forth. Let us not dwell whether this wine and spirit and lover is real or virtual, carnal or mystical. Even if the intent is mystical wine and love, it begs attention that the most prominent symbols of this world of muses and occult meanings is wine and love, two things that most numb the senses...the act of desiring that which sedates is proof of profundity of pain.[43]

Another attribute is a certain disregard, if not contempt, for native and non-native orientalists blinded, according to him, by the myth of the grandeur of the past and mystical character of the East. The solution according to him is not praising the past, but constructing the future by reviving the culture of industriousness which is universal, knows neither historical boundaries (past, present) nor geographical (East, West):

They [native idolizers of ancient Persia] want to excavate *Zend* and *Avesta* from the depths, with electric drills. If only we excavate our mines with the same vigor, instead of our superstitions, we would certainly become a first-rate nation.[44]

We too can attain the level of European civilization, though it may appear as a fantasy. However, if we want to persist in ways of the *Sufi* and recluse, uttering of wine and wine-tavern, we have to admit in all fairness that we ought not be entitled to dwell in the civilized world and should wither so those more industrious take our place.[45]

In Persia, the norm in scholarship is research on panegyric poets, convoluted prose interspersed with long-winded phraseology, translation of the most futile works from Arabic, mimicking the habits of retrograde researchers in Europe (Gustave Le Bon, *La Civilisation Arabe*, with essentialist view of race). It would not be regretful if our native scholars did solid research. But so far, they have not been able to produce a single work that can match the contributions of fine European scholars like Markwart's

Eranshahr, Christensen's *Sassanid*, Nöldeke's *Ferdowsi*, and to leave behind something truly worthwhile.[46]

Finally, he takes aim at scholars in Europe who seek Eastern potions for Western predicaments:

> What he (Henri Bergson) considers the epitome of mysticism, idle poets of the East have said as much in medieval times. This word love, more love, again love, has been said and written about so profusely and has been hoisted to an object of idolatry, the ring of which is deeply repugnant to a rational thinker.[47]
>
> Modern society has reached a dead-end, such that if it stays unchanged, it is fatal to human civilization. Depressed economy and unemployment run counter to technical inventions; social devastation runs counter to artistic creation. Industrial activity in recent years has been exclusively for purposes of destruction and war; a new foundation must be laid.[48]

We sense that he did not find a great deal of inspiration in the mainstream works of orientalism, barring certain works of scholarship (anchored in holistic treatment of cultural history), as there is little in there relevant to the challenges of modernity, both for the East and the West. Rather, it seems orientalism or its dominant strand is about reconciling the soul with predicaments that seem insurmountable; fatalism cloaked in scholarship, which he found distasteful.

The field of psychology, however, could address these shortcomings in a fundamental way. In his view, psychology is not only a complement to the natural sciences (mind probing the mind as a complement to mind probing nature), but it is also the foundation of human sciences (mental culture as a complement to material culture). And mental culture is a reflection of man's neurobiological traits, individual psychology aggregated at the societal level. Therefore, it is to psychology that he turned, with verve and intensity, ironically as this field was experiencing turmoil brought about by the Great War. Was not this catastrophe proof of the ill-fate of the human race, its mental apparatus predisposed to self-destruction? Taghi Erani would gravitate toward a contrarian view, inspired by Wilhelm Wundt, and his school of thought in individual and cultural psychology, which was drifting to oblivion in the aftermath of the Great War. But for him that is where the fundamental answers lay; answers that would dictate his personal conduct and ultimately his fate.

NOTES

1. Raziollah Hakim-Elahi and Akbar Afshar-Qotooli were two print-setters in the Erani circle, arrested with the "fifty-three"; Khame'i, *Panjah Nafar va Se Nafar*, pp. 76–77, describes the latter as a devout disciple (Fada'i) of Erani.

2. See Afshar and Dehbashi, *Khaterat'e Jamalzadeh*, p. 123: "Typesetter of the print-shop was a German and did not know Persian and we were obliged to correct the samples six or seven times which was quite taxing."

3. See Erani, *Tadghighat'e Lesani*.

4. See Erani, *Resalat fi Sharh ma Ashkal men Mosaderat Kitab Eqlidos*, pp. I–XIX.

5. Scribe was Mohammad bin Badr al-Ja'jarmi; manuscript was from 741 AH (1340–1341 AD).

6. See Erani, *Resalat fi Sharh ma Ashkal men Mosaderat Kitab Eqlidos*, pp. I–XIX.

7. See Frederic Rosen, *The Algebra of Mohammed Ben Musa Khuwarizmi*.

8. See Levenson, *Einstein in Berlin*, p. 91.

9. See *Bankverbindung*, p. 149 (recto–verso).

10. See Erani, *Badaye'e Afsah'ol Motekale'min Sheikh Muslih'ol-Din Saadi Shirazi*, preface.

11. See Machiavelli, *The Prince*, p. 13.

12. See Zakani, *Mouse and Cat*, pp. 75–76.

13. See Saadi, *Golistan*, p. 23: "Well-intentioned falsehood better than mischief-exciting truth."

14. See Saadi, *Bustan*, pp. 133–134.

15. See Saadi, *Golistan*, p. 62.

16. See Saadi, *Bustan*, p. 39, and *Golistan*, p. 116.

17. See Saadi, *Bustan*, p. 8, and *Golistan*, p. 86.

18. See Saadi, *Golistan*, p. 54.

19. See Erani, *Badaye'e Afsah'ol Motekale'min Sheikh Muslih'ol-Din Saadi Shirazi*, p. 6.

20. See Browne, *A Literary History of Persia*, Vol. II, p. 533.

21. See Erani, *Montakhab Lataef'e Molana Obeyd'e Zakani*.

22. See Lange and Mecit, *The Seljuqs*.

23. See Erani, *Ketab'e Vajh'e Din, Hakim Nasser-Khosrow*.

24. See *Bankverbindung*, p. 56 (recto–verso), p. 57.

25. Ibid.

26. Ibid.

27. Ibid., p. 58.

28. Ibid., p. 55 (recto–verso).

29. Ibid., p. 61 (recto–verso).

30. Ibid., p. 149 (recto–verso).
31. Ibid., p. 150 (recto–verso). Also see *Verzeichnis-Vorlesungen* and *Verzeichnis-Personal.*
32. Ibid., p. 306 (recto–verso), p. 307.
33. Ibid., p. 304.
34. Ibid., p. 302 (recto–verso), p. 308 (recto–verso).
35. Ibid., p. 302 (recto–verso).
36. Ibid., p. 103.
37. See Erani, *Resalat fi Sharh ma Ashkal men Mosaderat Kitab Eqlidos.*
38. See Hermann Weyl's *Raum, Zeit, Materie* (Space, Time, Matter); also Erani, *Farziye'e Nesbi.*
39. See Erani, *Donya,* 12-issue Volume, pp. 220.
40. Ibid., pp. 218–220.
41. Ibid., p. 110.
42. Ibid., pp. 24–25.
43. Ibid., p. 367.
44. Ibid., p. 375.
45. See Erani, *Teknik, Fann va Farhang* (excerpts in Behnam, *Berliniha,* pp. 194–195).
46. See Erani, *Donya,* 12-issue Volume, p. 169 (footnote); and pp. 2, 260 on Gustave Le Bon.
47. Ibid., p. 122 (elaborated in pp. 120–124).
48. Ibid., p. 309 (remarks appearing as footer beneath the two-column format).

CHAPTER 8

Prewar & Postwar Psychology

The main clue we find as to how Taghi Erani may have been exposed to psychology in Berlin is the *Lebenslauf* or vitae of his doctoral dissertation. There we find the name of Professor Koehler, and the latter (on the authority of Professor Wertheimer,[1] historian of psychology) is in all certainty Wolfgang Köhler, one of the chief advocates of *Gestalt* psychology, with its accent on the holistic nature of perception.

Köhler was the director of the Psychological Institute at the University of Berlin (1921–1935) and was a former student of Max Planck and Carl Stumpf. As a student, he had studied a domain straddling between physics and psychology, perception as triggered by acoustic phenomena. This later led to allegations that he had been involved in eavesdropping operations in Tenerife Islands during the war (connected to movement of British fleet and German U-boats), where he was conducting research on the problem-solving behavior of primates (chimpanzees).

Stumpf, his doctoral advisor, was also a proponent of *Gestalt*, and after a distinguished career in Göttingen and Würzburg had founded the Psychological Institute in Berlin and led it for nearly three decades (1894–1921). There was, therefore, a strong tradition of psychological research at Berlin. However, it was not on a par with that of Leipzig, which had been founded by Wilhelm Wundt in the 1870s, and had become the Mecca of physiological psychology, and its evolution to cultural psychology (*Völkerpsychologie*). Taghi Erani leaned towards the teachings of Wundt. Likely, Taghi Erani had a predisposition to delve

© The Author(s) 2019
Y. Jalali, *Taghi Erani, a Polymath in Interwar Berlin*,
https://doi.org/10.1007/978-3-319-97837-6_8

into psychology by virtue of his medical training in Persia. His curriculum included not only modern but also traditional medicine. Graduates had to be able to function in Persia, with a penury of modern drugs and implements needed for diagnosis and treatment, especially after the war. In the traditional curriculum, psychology figured prominently; that is, psychology as it had been handed down from the famed physicians of medieval times (Razi, Al-Balkhi, Avicenna, all with treatises on psychology).

Further, among the professors teaching natural sciences at *Dar-ol-Fonun*, in the high school curriculum, there was an instructor (Ali-Akbar Siassi) who had studied psychology in France before and after WWI, within the tradition of Théodule-Armand Ribot and Jules Payot.[2] There was also Persia's Minister of Science (Khalil Saqafi), the superior of *Dar-ol-Fonun* director, who had studied hypnotherapy in Paris in 1890s, under the neurologist Jules Bernard Luys.[3] Consequently, an infiltration of modern psychology was beginning to take place and find its way into academic circles in Persia.

The best way to develop an appreciation of the thought universe of Taghi Erani in the realm of psychology, is to get a sense of the psychology of Wilhelm Wundt. However, the irony is that by the time Taghi Erani had arrived in Berlin, Wundt had not only passed away (1920), but his school of psychology had fallen by the wayside. That is not to say that something more substantive had taken its place, but that there was a Balkanization of the field, and only those schools that had roaring guardians seem to have managed to garner a certain following.

In America, the behaviorist school won the day (Watson its champion and later Skinner), a situation that did not change until the rise of cognitive science in 1950s; the latter with an emphasis on the intrinsic structure of the brain and the nervous system, and the generative characteristics of the mind, largely ignored by the behaviorist school.

The behaviorist school was also favored in the Soviet Union, which concocted its own brand called "reactology," championed by Konstantin Kornilov with focus on determining the limits to which the mind could be reengineered, by observing how the subject reacted to engineered situations.

In the tug of war between the introspective school (founded by Wundt) and behaviorist school (with an emphasis on mapping stimuli to response), it was the latter that gained ascendancy by drawing on the arsenal of Pavlovian conditioning and British associationism.

Meanwhile, another tradition began to impose its presence, that of meta-psychology, which today is better known as psychoanalysis. The dominant variant of this school maintained that what governs man's behavior is the resultant of its animal instincts and socially imposed ethos, causing perpetual tension and possibility of pathogenesis. This propensity was in some sense an invariance of the human condition, with which one could cope but never overcome or resolve.

In contrast to this, Wundt drew on anthropological research to artic-ulate the notion of mutability and plasticity of mentality and behavior governed by volitional processes, derived from man's biological deter-minants (physiological and evolutionary), psychological endowment, and sociohistorical milieu. He developed an elaborate theory of the growth of mentality over the course of known human history, anchored in anthropological evidence, highlighting its contingent and complex nature.

PSYCHOLOGY, MEANS AND MORALE, DEATH OF OPTIMISM

What a group of people can achieve, as conventional wisdom would sug-gest, depends to a large extent on their spirit and morale. What is some-times missing when we examine the history of a discipline is clues about why in a particular period there is so much achievement and progress, while in others, perhaps longer in duration, there is scant movement and production. What it takes to develop a better picture, therefore, is not only the catalogue of results, but also the means mobilized for its achievement and the spirit governing the whole enterprise.

In terms of means, psychology had no reason to be envious of other fields, particularly during the period of its rapid growth (1870s–1920s), with the exception of the fundamental sciences—physics, chemistry, biol-ogy, mathematics—on which a great deal was riding, in existential terms, global competitiveness.

Notwithstanding this bias, psychology was not doing too badly in the grand scheme of things. In the USA, its appeal to business for stimulat-ing worker productivity and in education to stimulate the learning pro-cess gave it a good shot in the arm. In the Soviet Union, it was treated as strategic science vital to the state. And in Europe, it got its fair share. When Stumpf arrived in Berlin in 1894 he had only three rooms to cre-ate his Psychological Institute. By the time he retired in 1921, this had grown to twenty five, with commensurate increase in staff and budget.

Resource of course is one thing, and spirit another. While spirits were running high, perhaps too high, before the war, after the war it was a different story; and not only in Germany, but across the board. For one thing, the war wiped off the camp of those we might call the "optimists." William James passed away before the war (1910), and that was perhaps best, for he had a most gentle personality, and no telling what this tragedy would have done to his psyche. Already in the 1890s he was preaching how there is no single vision of what is right or wrong, and all have a right to think and behave according to their culture and traditions, and the best philosophy is to live and let live. He called it pragmatism, in the sense that only that which yields peace and tranquility can lay claim to validity and truth.

James was at odds with the pessimists, whose condition he attributed to over-education and having lost touch with the most basic of human instincts, to live for the moment, to not intellectualize, to find joy in simple things and sensations. He regarded Walt Whitman, the poet, as a modern-day prophet, but also with the warning that you cannot be a prophet and a man of the world at one and the same time. He was fond of Herbert Spencer who believed in the universal laws of evolution carrying us forth toward ever greater perfection and completeness.

But optimism had always been a minority voice in the humanities—philosophy, psychology, literature included, except during the brief period of the Enlightenment (eighteenth century), when reason was poised to conquer unreason and propel us toward a brighter future.

In antiquity, the moral philosophers of Greece, those contemplating human and social predicaments as opposed to natural phenomena (Socrates onwards), all came in the wake of Draco and Solon, statesmen who scrambled to save Athenian society from devouring itself. Debt had led to enslavement of freemen, its eradication indispensable to save society from imploding, "Draconian" laws notably against usury. When the house is on fire, people start thinking of how to quench it, and the thinking is more akin to panic and improvisation than hope and optimism. John Dewey, who was a kindred spirit of James, couched this concept in technical terms—when one's desires are unmet, thinking occurs, and when fulfilled, it ceases.[4]

Neither was optimism in vogue during the Middle Ages, which posterity has come to call the Dark Ages, not even in the Islamic World, which was thriving commercially, but remained austere intellectually. The dominant figure, Al-Ghazali, shattered the clockmaker hypothesis (an attempt by the rationalists to reduce God to the act of creation and have man take over thereafter), articulating a dynamic God who remained engaged throughout the process of unfolding of nature, with little for man to do, but abide by the role assigned to him.[5] And the best the *Sufis* could do to counter this was pantheism. That God is nothing but nature, but man still kept a submissive role with respect to nature, trying at best to reunify with it, as opposed to shaping it, with no room for optimism.

As for modernity, recent scholarship[6] has discovered that even Leibniz who was mocked by the skeptics for his philosophy of optimism (Voltaire's jabs—the universe is so finely designed that we even have a nose on which can rest our spectacles for the benefit of our eyes!), had repudiated his famous formula that *"we live in the best of all possible worlds."* Only that Leibniz left his latest schema of the conception of the universe in his drawers, which saw publication two centuries after his death, proving that the great thinker had come to more sober thoughts about the nature of our existence.

In summary, optimism was never a dominant strand in the humanities, and the Great War made sure that it would never become one, and in fact, would cease to exist. A century separates us from that period and it can still be glimpsed in literature, but as a pleasantry. Chekhov began this before the war—*"After two or three hundred years, life on this earth will be unimaginably beautiful and extraordinary"*—such is Vershinin's wishful thinking in *Three Sisters*; and Beckett elevated it to an art form. Only simpletons hold out for their wishes to be fulfilled (*Waiting for Godot*).

REIGN OF PESSIMISM

Given this state of affairs, the pessimists were emboldened beyond measure, and as if to give proof to Aristotle's law of nature that it abhors emptiness, they immediately filled the space vacated by the cadaver of the defunct creed (optimism) that had just been carried off the world stage. Here, we are confronted by an *embarrassment of riches*. By examining two characters, one reflecting the intellectual ambience of interwar years, especially in Germany, the other from the ranks of psychology, also from the German-speaking world, we can catch a glimpse of this philosophy.

It must be said beforehand, though, that the pessimists had a lot going for them—one was the stark reality that had just come to pass, the devouring of *one percent* of the global population at the drop of a hat; a little warning that worse was in store if man did not behave. And why would man behave any differently now than it had since antediluvian times. Can humanity change its behavior, for it stems from its nature; can humanity change its nature? (In WWII the figure was *three percent*, a fourfold increase in the number of fatalities, reflecting the increased world population.) The other, was the intellectual and literary prowess of the champions of pessimism, some the best in the business of capturing the imagination of the public, and of the intellectuals. Finally, and most importantly, was the demoralized psyche of the audience to which they preached. Often, what carries the argument is not so much the force of the argument as the feebleness of the audience; and the feebleness of the audience has less to do with its intellectual faculties than its morale and spirit. In the aftermath of the Great War, morale was at rock bottom, just as feebleness was at its zenith.

***

Spengler coined the term "morphology of history," and on the European horizon wherever he looked he could detect an atrophied morphology growing like cancer (he puts the frontier of descent into hell squarely at year 1800)—in art, literature, architecture, science, religion, philosophy, the whole lot.[7]

The artist is no longer a creator but a mere craftsman, so much so that not a thing will happen to the development of art if all schools of art were to shut down immediately; the writer, a producer of prose, subscribing to pedantic rules of style instead of bursting with passion and provocative ideas; architecture devoid of vivacity, no longer a celebration of creativity and imagination, but a demonstration of austerity and authority. Similarly, the language of science is only about universality, laws, force, power, energy, rather than proportion, perspective and harmony. And atheism is the new religion, not as a creative conception but as the exhaustion of all religious possibilities; and philosophy, systems with scaffolding of steel, crushing all dissenting views under its deadly weight.

In short, it is a culture devoid of soul, spirit, spontaneity; more precisely, not culture but civilization for the former implies generative qualities, the latter degeneration, and its greatest achievement—the Megapolis, the World-City (London, Paris, Berlin) in which dwell soulless creatures, bouncing to and fro, gazing at the austere architecture with its intimidating presence, engaged in science, art, philosophy, ordinary life, but of lifelessness, dullness, nothingness; hell on earth, without any escape.

Spengler was a big hit, for he gave to the bruised psyche of the post-war European the peace of mind that there was nothing he could have done to forestall the catastrophe. It was bound to happen, and it happened. History abiding to the laws of nature—growth, maturation, decline, collapse. An enormous psychological relief, one may even say liberating, a great bargain for toiling through a tome of doom and gloom.

Freud did not care to make a distinction between culture and civilization[8]—he said so explicitly, no doubt alluding to Spengler. For him, the whole thing was rotten, culture and civilization confounded. He would outdo Spengler, and in true Freudian style.

Man had come into existence as a free being and only through coercion, had resigned himself to collective life, gaining security in the bargain; but also submitting to the dictates of the group, or rather the minority who had conceived it. And this authority came with suppression of his innermost instincts. Through millennia, this tyranny was internalized into feelings of shame, guilt, sinfulness, and what came to acquire the technical term—superego. For ego was just the self, living between the fires of man's animal instincts, the id, and the oppression of group expectation, the superego, the two colluding to crush the ego, which longs for its long-lost individuality—going back to the state of nature, to idling and not laboring if nourishment is at hand, to pursuing his animal instincts of copulation, aggression, cruelty and annihilation, of others and possibly of the self.[9]

Freud even discovered a new instinct, the death instinct, that was so insidious that even the ablest of psychoanalysts, with the exception of Adler, could not see it. But Freud could see it, as it was just shadowing the life instinct. Didn't certain animals die upon copulation; that in itself is the strongest indication of the infinitesimal proximity of life and death instincts. Didn't love turn into hate as if by a wand, hate into a death wish; a death wish into an act of crime, in a moment of madness perhaps, but recurring with daunting and damning regularity. Certainly, this contains a kernel of man's innermost nature and constitution; it's not a fluke or coincidence.[10]

There is no escape from this inferno. There is a solution but one which is impossible to implement. Only if everyone agreed to live collectively under different rules, not under the hypocrisy of religion—which is nothing but the codification of the superego, institutionalized illusion, bordering on delusion (and he is solely concerned with the "European

Christian Civilization" which had just engaged in a barbaric act of self-mutilation)—but under the banner of science which recognizes the true and real nature of man. Hence, this would allow him to live in a state of seminature but amongst men, to be guarded from the vagaries of nature—perhaps then there would be a glimmer of hope. But that would require that man act out of reason and not out of passion. Given he has the highest propensity to act out of passion and instinct, and not reason, why would an awakening and rebirth happen now. "Perhaps I too am chasing an illusion," so concludes Freud.[11]

Freud may have had his reasons to feel this way. His professional followers were falling off, first Adler, then Jung, and others also, as they felt he had overstated his case to the detriment of psychoanalysis, the discipline he had created nearly single-handedly, at least so he claimed.[12] Then came the war and his favorite daughter died of disease spread by the war. Then, he lost his only connection to her, her son, who died a young boy. Then came the tide of anti-Jewish sentiment in Austria, following the footsteps of Germany. This was followed by the burning of his books in Germany and the annexation of Austria by Germany, causing his flight from his beloved Vienna. Finally, he succumbed to the cancer from which he had been suffering for years (malignancy of mouth tissue, caused by his cigar-smoking habit), within eighteen months of arriving in the UK, only weeks after the start of WWII. For him, his prophecy had come true. Man was destined to self-destruct.

Hence, for psychology, the paradox was that just when it had come to acquire the means necessary to achieve significant results (1920s), it had lost its spirit, in contrast to its earliest days (1870s) when it was overflowing with enthusiasm and ideas, but barely had the means to carry out a single experiment. It would take the curiosity of an open-minded physicist, Helmholtz, to scratch the surface of that infinitely complex thing—man's mental life—by experimentation and observation, as in those early days there was not much else.

After the 1920s, psychology grew enormously in size, but in terms of theoretical substance there was little that was added (insights from *Gestalt* on holistic perception), a situation that did not change until the 1950s and the advent of cognitive science. This physical growth, associated with the professionalization of the discipline and its penetration into a wide range of applications (military, marketing, media, amongst others), disguised the lethargy of psychology in terms of its scientific substance. It was somewhat like a baby growing in size but not growing intellectually, and if anything, showing occasional signs of regression.

The man most recognizably associated with the ascending arc of psychology, a half century of solid progress (1870–1920), before it went into the doldrums (1920s–1950s), in a sense the golden age of psychology, and an important influence on Taghi Erani, was Wilhelm Maximilian Wundt.

WILHELM WUNDT, VOICE OF REALISM

Wundt founded the field of physiological psychology, rescuing it from abstract philosophical speculation, and mechanistic physics, penetrating the processes operating in the mind as an agent with inherent activity, stressing that the mind is a property of the brain and the nervous system, attached to an organism which is bound to physiological and evolutionary laws of behavior, with a history programmed in reflexes, instincts, memory, thought, language. This signified that a variety of behaviors could be observed from a similar set of circumstances, as the organism evolves and adapts to its physical and social environment, driven by instincts of satisfaction, survival, and procreation.

It was a rich and dynamic world that Wundt depicted, and he developed a comprehensive lexicon, concepts, measurement methods, interpretation techniques, and observational principles to give structure and solidity to psychology as an autonomous science, with connections to other sciences, physical sciences and historical sciences.

He established Leipzig as the Mecca of experimental psychology, supervised nearly two hundred doctoral dissertations, students (including one of the earliest female psychologists, Anna Meyer Berliner) who went on to establish departments of psychology across Europe and USA. He ventured into the field of *Völkerpsychologie* (now called cultural or ethnic psychology), demonstrating an encyclopedic command of the anthropological field work and knowledge of his time, enabling him to outline the main stages of maturation of man's mentality through the ages.

Wundt was disciplined in separating science from its application. His mission was to establish the science of psychology as he saw it. It would be for others to decide how they wish to apply this knowledge toward good or evil, mundane or noble goals, immediate or long-term objectives. Those who cannot make this distinction are prone to producing a muddled science, rationalization of their *a priori* biases disguised as science.

Wundt was not averse to addressing philosophical questions from the perspective of psychology, such as the notion of free will. He thought James's formulation was naïve from a psychological point of view;[13] nor was he averse to describing drives, ego, inner motives, waxing and waning of consciousness, the realm of psychoanalysis, but does not seem to have commented on Freud. Given what he thought of James (fine writer, but not a contributor to psychological knowledge[13]), it is conceivable that for him Freud was the practitioner of a healing art, setting forth hypotheses for his co-practitioners as manuals; not relevant to psychology as science, as Wundt conceived it.

Below, passages from Wundt concerning his concept of volition, central to his psychology, and concepts in folk psychology and its bearing on modern civilization, which give us a glimpse of his thought universe:

> Great social transformations, of which one of the greatest is unquestionably the transition from the primitive horde to totemic tribal organization, are never affected by the ordinances of individuals, but develop of themselves through a necessity, immanent in the cultural conditions. Their effects are never foreseen but are recognized in their full import only after they have taken place.[14]

> His [primitive man's] morality is dependent upon the environment in which he lives. Where he lives his life of freedom, one might almost call his state ideal, there being few motives to immoral conduct in our sense of the word. On the other hand, whenever primitive man is hunted down and hard pressed, he possesses no moral principles whatsoever. These traits are worth noting, if only because they show the tremendous influence which external life exerts, even under the simplest conditions, upon the development of the moral nature.[15]

> The technical advance of civilization carries with it serious danger of a vast moral retrogression...But it would be preposterous, again, to try and save the situation by any vain effort to reverse the wheel of the world's progress. The dangers that come with civilization can be met only by the further advance of civilization.[16]

Notwithstanding the richness of Wundt's psychology, the student of interwar years, certainly in America, which increasingly occupied center stage as Europe sank into war, was in no mood to plough through his vast corpus, preferring at best secondary accounts which tended to dilute if not misrepresent his school of thought (notably misrepresentation by Titchener and Boring, as historians of psychology have established).[17]

Thus, slowly but surely Wundt eclipsed to oblivion. Only since the 1970s, has there been a revival of interest in Wundt and a return to the original works to illuminate the substance of his ideas.[18]

***

Wundt had a profound influence on Taghi Erani, judging from his accent on neurobiology, affective processes, volition as the central problem of individual psychology, historical psychology as a means to understand prevailing mentality and ethics.[19] He acknowledges this somewhat, but partly it is only implicit in his analyses, as he was juggling other influences in articulating his own views.

Taghi Erani was also influenced by Freud, but Freud of the prewar period (when he laid the foundation of his school of psychoanalysis), judging from his commentaries. It is questionable if he had ploughed through Freud's works that were produced after the war, which veered toward what we may call societal and civilizational psychology, which tended to be rather dark.[20]

There is, however, a convergence between Taghi Erani's notion of social neurosis, developed in *Mysticism and Materialism* (1934) and concepts in Freud's *Civilization and Its Discontents* (1930). Briefly, it can be said that the convergence is in terms of recognition of neurosis manifested at societal scale, and it does not carry over into its specific form or the analysis of root cause.

On that score, Erani seems to be neither with Freud nor with Wundt. Perhaps in keeping with his youthful spirit, he proceeds to the stage of diagnosis (specificity of the neurosis) and treatment plan, social action, shaping the subsequent phase of his life. We now delve into this psychological universe of Taghi Erani.

NOTES

1. Private correspondence with Prof. Michael Wertheimer.
2. On Ali-Akbar Siassi, see Hashemian, *Tahavolat'e Farhangi'e Iran*, pp. 167, 470–471, also Schayegh, *Who Is Knowledgeable Is Strong*, p. 159.
3. On Khalil Saqafi, see Mohammadi-Doostdar, *Fantasies of Reason*, pp. 195–204.
4. See Wertheimer, *A Brief History of Psychology*, p. 106.
5. See Leaman, *A Brief Introduction to Islamic Philosophy*, p. 25.
6. See Russell, *A History of Western Philosophy*, pp. 590–591.

7. I have relied on the abridged version of Spengler's *Decline*, pp. 4–57.

8. See Freud, *The Future of an Illusion*, p. 263.

9. Ibid., pp. 264, 266; also see *The Ego and the Id*, pp. 242, 244.

10. Adler's concept of aggression drive (frustration of urge toward fulfillment of one's desires and potential), came years before Freud's death instinct, and is considered to be its forerunner. On Freud's notion of death instinct, see *The Ego and the Id*, pp. 246–247, 250, 255.

11. See Freud, *The Future of an Illusion*, pp. 279, 281, 283, 285, 288, 289.

12. Ibid., p. 282.

13. See Rieber (ed.), *Wilhelm Wundt and the Making of a Scientific Psychology*, p. 129 (chapter by Arthur Blumenthal).

14. Wundt, *Elements of Folk Psychology*, p. 152.

15. Ibid., pp. 114–115.

16. Wundt, *Ethics*, p. 315.

17. See Rieber (ed.), *Wilhelm Wundt and the Making of a Scientific Psychology*, pp. 73–87 (chapter by Kurt Danziger).

18. Ibid; contributors to this volume have produced works to revive and re-examine the psychology of Wundt:

    P. 134: Blumenthal, A. L.: "A Reappraisal of Wilhelm Wundt," *American Psychologist*, 1975, 30, 1081–1086.

    P. 134: Blumenthal, A. L.: "The Founding Father We Never Knew," *Contemporary Psychology*, 1979, 24, 449–453.

    P. 87: Danziger, K.: "The History of Introspection Reconsidered," *Journal of the History of the Behavioral Sciences*, in press (1980).

19. Influence of Wundt's psychology on the psychology of Taghi Erani—emphasis on evolutionary biology, physiology, neurobiology and affective motives and drives in shaping mentality and behavior; psychology as the science of the mind (internally focused); mind as possessing its laws of causality, in constant activity and inventiveness, even within (hypothetically) non-changing environment; volition as the central theme of personal psychology; refutation of free will (as opposed to physical–psychical determinism of volitional processes); evolution as a process of splintering of complexes into differentiated strands, recombination, constant regeneration; growth of thought along path of concrete to abstract; evolution as genealogical growth (new form appearing in existing form, continually); historical psychology as the most important domain of psychology from the perspective of understanding prevailing mentality; social psychology as a slice of historical psychology; spontaneity of formation of primitive societies and comprehension of its consequences in retrospect (vs. social contract or coercion theory of formation of primitive societies articulated by Freud); production of magic, myth, religion, philosophy, science, art and aesthetics as genealogical process of one form flowing out

of another (continuously discontinuous) and as collective-social enterprise; irreversibility of the sociohistorical process and its contingent character (Wundt's law of heterogony of ends, Erani's dialectics of change antithetical to teleology); human volition and its evolution the determinant of historical contingency.

20. Freud's Works after WWI: *Beyond the Pleasure Principle (1920), The Ego and the Id (1923), The Future of an Illusion (1927), Civilization and Its Discontents (1930), Moses and Monotheism (1939).*

CHAPTER 9

Autodidact in Psychology

We recall that Taghi Erani began work at *Kaviani Press* as a proofreader shortly after his arrival in Berlin, in September 1922. *Leibnizstrasse 43*, in the leafy suburb of Charlottenburg in west Berlin became his moonlighting den, while during the daytime he was engaged in his studies in chemistry and laboratory work at the university.

During 1924–1925, he annotated and published three works from literary–philosophical figures of medieval Persia (Nasser-Khosrow, Saadi, Obeyd Zakani), also at *Kaviani Press*. Then, he shifted gear to his project on the *Series in Exact Sciences*. The scope of this project consisted of condensed tomes on physics, chemistry, biology, psychology, and logic. The plan for this series, as well as its first installment (*The First Volume of Physics*) was published in January 1926 by *Kaviani Press*.

One of Erani's motives for working at the press was to facilitate self-publishing. Initially, the motive had been financial; but after he landed the job at the Oriental Languages Institute, where he taught from May 1925 to December 1928 (his departure from Berlin), he had a reasonable income.[1] Additionally, during the last two years in Berlin, he received a stipend from the government of Persia.[2]

In December 1927, Taghi Erani published a major work on psychology, again with *Kaviani Press*. This work was later expanded when he returned to Persia and was published in May 1932 by *Sirousse Press*. The main addition was a section on sociohistorical psychology.

© The Author(s) 2019
Y. Jalali, *Taghi Erani, a Polymath in Interwar Berlin*,
https://doi.org/10.1007/978-3-319-97837-6_9

In 1934–1935, he published a number of essays, serialized in his journal *Donya* (*Le Monde*), in several of which there are forays in psychology. At times, it is difficult to categorize these works. For instance, *Mysticism and Materialism* (1934) traces thought movements with a particular psychological disposition in Eastern and Western thought; he covers China, India, Persia, Greece, and Europe from antiquity to modern times, touching on religion, philosophy, and science. This work could be interpreted as a narrative of a psychological phenomenon characterized by introspection, or history of a philosophical school of thought. The lines cannot be so sharply drawn.

Nevertheless, to assess his work in psychology, it cannot be limited to the single book he produced on psychology. For instance, in getting a sense of his views on Freud, which is largely positive, corresponding to Freud's works prior to the war where he laid the foundations of his school of psychoanalysis, we have to collate Taghi Erani's remarks from his book on psychology and four other essays; below, excerpts on Freud, scattered in his writings.

Freud's theory in the psychology of sexuality is no doubt the most significant application of dialectics in psychology, on a par with Darwin's theory of evolution in biology, the materialist interpretation of history, and the mutability of space-time in the theory of relativity. According to this theory, two instincts, the urge for nourishment and the urge for procreation are the primordial drives of the living organism. The drive to procreate loses its semblance to biological instinct in the bosom of society, asserting itself in disguised and sublimated form.[3]

Disorder in sexual function (Sadism, Masochism, excesses, deficiencies and abnormalities) may be caused by hereditary factors or neurophysiological defects from birth....(there are psycho-medical procedures for coping with these)...Freud is of the view that acts and thoughts of sexuality arising in childhood are suppressed in our thoughts [by social mores], and this gives rise to ill-thoughts and illusions, the principal forms of which have to be singled out, in order to stimulate the patient to be extricated from their hold. There is opposition to Freud's views as his therapeutic methods have not yielded positive results in consistent manner.[4]

If we wish to rationalize sleep and dreams from the perspective of psychology and psychoanalysis, it has to be said that humans by nature are not keen on toil and labor, and wish to be free of its burden. Children are not fed up with life, for this reason they have no desire to go to sleep.... Freud is of the view that distress in waking life motivates sleep....dreams relate to past happenings of our life that have become disguised beyond cognition and comprehension. In dreams, we relive those experiences.... mental phenomena that occur in waking life, leave a trace in our memory; connections may exist between different memories and produce various representations. In sleep, internal factors (sexual excitation, fear), and factors of the environment may trigger resurrection of these representations, which we call dream content. Therefore, dream is a consequence of experienced events...From the nature of the dream the mental life of the dreamer can be explored.[5]

In order to retain certain things vividly, memory demotes other things, causing forgetfulness. This may be superficial and temporary or deep and permanent...Freud is of the opinion that the individual loses memory of unpleasant things, by a process that is subconscious.[6]

Freud says that to proceed with the treatment of neurosis, the neurotic condition must first be acknowledged by the patient. That is, the patient must become cognizant of his or her condition. This principle also holds for society at large. Recognition of defect of society is the first step towards its treatment. Just as celestial creatures that were inventions of man have been dethroned by man, the hold of society over man, which is man's own creation, will also be dethroned.[7]

BIRD'S-EYE VIEW OF TAGHI ERANI'S PRINCIPAL WORK ON PSYCHOLOGY

Taghi Erani's *Psychology* is a condensed work three hundred pages in length, written in Persian, first published in Berlin in December 1927 by *Kaviani Press*. It begins by framing the subject matter and terminology, moves to neurobiology as the basis and incubator of mental processes, elaborates on mental processes operative in individual psychology, and sketches mental processes operative in human society and history: the latter a succinct survey of magic, myth, religion, philosophy, science, logic, and ethics.

The work is titled *Principles of the Science of the Mind, General Psychology (Osool'e Elm'e Rooh, Pesikologi'e Omoomi)*. It consists of— overview (30 pages); anatomy and physiology of the nervous system (40 pages); connection of the mind with the external world (110 pages); internalization of external phenomena by the mind (60 pages); manifestations of the mind in the realm of individual psychology (30 pages); and manifestations of the mind in the realm of sociohistorical psychology (30 pages).

Therefore, its scope is the discipline of psychology in its broad contour, not a specialized account (e.g., child psychology, the focus and success of Piaget and Vygotsky, his contemporaries). Its aim is to demystify the workings of the mind based on the latest scientific theories and hypotheses, as focusing on the mind is the most direct path to developing insights about what governs human behavior. For this reason, psychology is not the science of behavior, if that means glossing over the inner workings of the mind. Rather psychology is the science of the mind which underpins behavior and illuminates it.

Further, in his lexicon "science" means explanatory theories which adhere to the laws of causality, without exception or exemption. However, causality is not only mechanical causality. There is also physiological causality, evolutionary causality, psychological causality, sociological causality, historical causality. For instance, a memory unleashing a chain of emotions and behavior is an example of psychological causality.

Therefore, the title of the work—*Principles of the Science of the Mind*—when viewed in the context of the intellectual climate of the time, is a manifesto of sorts. At that time, there was a marked shift away from psychology as the science of the mind to psychology as science of behavior (led by Watson in America, Kornilov in the Soviet Union, less pronounced but also visible in Europe).

If we had to characterize his approach—in the words of the late George Mandler, describing William James—we could say he is partly a summarizer, partly synthesizer, and partly theorizer.

In terms of intellectual indebtedness, we would have to place Wundt in the primary position, although the work is replete in references to a wide spectrum of thinkers and practitioners of the field, embracing philosophers, physicians, physiologists, physicists, chemists, naturalists, psychologists, psychiatrists, psychoanalysts, linguists, dramatists, and so on—Dante, Hobbes, Descartes, Huygens, Spinoza, Locke, Malebranche, Leibniz, Wolff, Reid, Hume, Rousseau, Kant, Pestalozzi, Goethe, Schiller, Dalton, Herbart, Chateaubriand, Wellington, Hegel, Arago, Weber, Fechner, Müller (Johannes), Darwin, Spencer, Virchow, Helmholtz, Müller (Max), Broca, Huxley, Wundt, Hering, Brentano, Bernheim, James, Nietzsche, Ostwald, Freud, Münsterberg, Stratton, Simon (Théodore), among others.

Nevertheless, he does highlight the prominence of Wundt when giving a recap of the development of psychology as an autonomous discipline:

Amongst natural sciences, initially physics had the most impact on the development of psychology. Then, discoveries in biology and chemistry lent support to this discipline. From 1830s onwards, sensory phenomena were investigated by researchers such as Johannes Müller, Weber, Helmholtz and others. For the first time, methods of experimentation and physical tests entered the science of psychology. Fechner demonstrated the use of mathematical formula; his method of psychophysics is well-known. Around 1860, time response of mental processes was measured; by 1870 it was concluded that there is no specific site in the brain corresponding to higher mental functions; rather such functions are distributed across the surface of the brain [cortex]. Finally, the German scholar Wundt compiled all the findings of his time and turned them into an organized body of knowledge which is the present-day science of psychology.[8]

There is a bent on historical–cultural melding of the mind in his work, which is thoroughly Wundtian. There is an accent on pedagogy and child development, reminiscent of Piaget and Vygotsky, whom he did not know (both started producing works in the 1920s, shortly before Taghi Erani). He knew of Théodore Simon's work, however, and describes his method in assessing the learning ability of children. Overall, there is a dominant strain of practical utility running through the work, reminiscent of James:

Psychology forms the fourth book in the *Series in Exact Sciences*. In physics, chemistry, and biology (analytical natural sciences), the mind investigates the processes in nature; in psychology, it investigates its own processes. For this reason, psychology complements natural sciences and forms the foundation of mental sciences...Pedagogy, medicine, law, social sciences, and other disciplines, are compelled to depend on this science. The teacher, physician, judge, public servant, parent, must have knowledge of the mental state of the pupil, patient, culprit, [citizen], and child, to be able to perform their duties diligently.[9]

Study of psychology can be likened to a school, wherein the objective is not to learn a specific domain of knowledge, but to learn how the learning process works. This subject has practical application. For instance, in our daily life we observe that understanding and retaining certain concepts may be easy, while others are difficult. No doubt this difference has a reason. We can investigate this reason with the help of psychology. Although psychology may not be able to facilitate the learning of various subjects to the same degree, even if we know the reasons, but developing this understanding is a big step towards penetrating the dark labyrinths of the mind.[10]

Neurobiology—The Foundation

The part on neurobiology covers neuroanatomy and neurophysiology. Erani, the former student of medicine, rejoiced in the prominent position that the brain and the nervous system play in the generation of the most basal to the most abstract products of the mind, and its impact on behavior, including internal behavior and emotions; drives, stirrings of the subconscious, and external behavior, including pathogenesis.

This was a time in the history of neurobiology when the first molecule (acetylcholine) acting as neurotransmitter had just been discovered (Loewi, 1923). We now know over one hundred such molecules. It was a period when it was still not a settled matter whether the nervous system is a continuous or disjointed network of neurons (Golgi-Cajal controversy, settled in the 1950s in favor of Cajal's disjointedness theory, by the use of electron microscope). This was a time before brain imaging with radioactive tracers and intrusive means (microelectrode) yielding resolutions of one millimeter and one microsecond was thought to be possible. It was a time with no inkling of chemical synapses, ionic gateways, retro-transmission (responsible for motor function without external stimuli, triggered by the favorite motives of psychoanalysts—drive, instinct, imagination, memory), and slow transmission (vital to learnability). Likewise, it predates our knowledge of ganglionic processing (streams of sensations summarized to abstract patterns prior to cerebral processing, vestige of an evolutionary past), the nature of the nerve–gland connection (making hormones big actors of instinctive and emotive behavior in which the limbic system plays a prominent role), and the dual character of visual perception in the cortex (vernal and dorsal streams specializing in object identification and localization).

Below are excerpts from Taghi Erani's *Psychology* pertaining to neurobiology:

> The function of the nervous system is the acquisition of information from the environment. This is because this function is vital to the preservation of life.[11] The nervous system consists of cells called nerve elements or neurons; neurons have common characteristics. Fibrils emanate from neurons; one type emanates from the cell protoplasm; these are called dendrites and

are short and numerous. The function of these is not well understood. It is only known that they do not regenerate. Another type is the axon, few but elongated; these lack chromatin and secrete a shiny fatty phosphorus substance called Myelin, which protects them as coating.[12]

Nerves have properties of excitation and transmission; impulses that cause excitation may be mechanical, chemical, physiological, or psychological; psychological impulses such as fear, volition, and resolve can cause excitation of the nerves.[13]

Mental phenomena, like phenomena in other sciences are contingent on time and space; that is no matter how fast they appear, they occur within a certain interval of time. Emotions are a consequence of physical, chemical and physiological processes triggered in the nervous system. Therefore emotions are contingent on space, although we cannot localize them...Force in mechanics might not have a specific locality either...nevertheless it is not possible for force to exist without space, as force is the stimulus that creates acceleration, acceleration is change of velocity, velocity is change in space....Likewise, if in the cells and molecules and atoms comprising the nervous system changes do not take place, mental processes do not materialize. We perceive mental processes of the self through internal sensation (introspection) and that of others through external manifestation (observation).[14]

If we remove the small brain (cerebellum) of a pigeon it will not be paralyzed, but its movements become disorganized; so, the role of cerebellum is organization of motor functions.[15]

The large brain (cerebrum) exterior is grey matter; its interior is white matter and is responsible for transmission of impulses. Philosophers of antiquity considered the cerebrum the seat of the soul. Seventeenth century rationalists imagined a fluid for the soul. Descartes thought this fluid exits the brain [pineal gland] and reaches the bodily organs through nerve tubes.[16]

The view that is widely accepted today is that there exists from birth an original endowment of notions of space and time (Kant), which gets refined as a consequence of sensory impingements.[17]

In contrast to speech, for which there is a specific site on the cerebrum (Broca), there is probably no specific site for higher functions like cognition, judgement, reasoning, logic; rather these result from the cooperation of different sites.[18]

Without language, precise thoughts as exhibited by man could not materialize. Speech, as distinct from language, is the means by which the individual exhibits the contents of the mind.[19]

Individual Psychology and Cultural Psychology

In *Psychology*, Taghi Erani describes the utility, scope, method, history, and main themes of individual and sociohistorical psychology. The latter begins with a description of development of man's mentality to the threshold of history, several millennia before the present. Then come complex thought systems of religion and edicts reflected in sacred and secular philological scriptures that spread from the East to the West, with ripples back to the East; but it is in the West that the mind reaches its culmination in modern scientific cosmogony. This is followed by the precipitation of a sense of crises, accentuated by the Great War and its aftermath and the germinations of a new beginning:

> Living organisms strive to acquire the means necessary for survival and deflect forces that jeopardize survival. To perform this function, the mind develops a particular structure and organization. It is the task of psychology to investigate this organization.[20]
>
> Just as the mind is an interconnected whole, so too is the science devoted to its study. When we divide psychology into separate spheres, it is only from the perspective of facilitating learning. As such principles of psychology can be elaborated only at the end of such piecewise exposition, not at the beginning. This makes it different from other analytical sciences where the general principles can be stated upfront, such as conservation principles. However, we will adopt both approaches to familiarize the reader with the convoluted topics of the mind. At the start, we describe the subject-matter of psychology, its outline and evolution as a science, and at the end elaborate the aggregate principles.[21]
>
> Primitive man experiments with numerous things; distinguishes beneficial and harmful plants; hunts animals to feed itself; protects itself from wild beasts and insidious creatures; uses fire, water, rocks, wood to maintain itself; learns how to smelt metals; counts things with its fingers; gauges distances with its hands and feet; gazes at the sky like a child; notices change and motion and in this way adds to its knowledge of the external world; these help it satisfy its vital necessities; if in these ventures it stumbles on some information, it uses that too to serve its needs. The accumulation of these findings is not yet science; man has not established the relationship between these phenomena; science appears when man's material necessities are somewhat secure, his mental faculties somewhat developed, causal relationships somewhat apparent; but material progress ushers the splintering of society into classes and factions; tension and conflict results; privileged factions strive to preserve their status; retrograde

thoughts proliferate; these persist until conservative factions and the mode of life on which they subsist withers.[22]

Primitive man's refuge in mythology is akin to a child stricken with awe. He takes refuge in spirits because of despair. Fear and need give rise to religion...through a long process totems and spirits are subsumed into a unitary god of the celestial sphere. Monotheistic religions such as Zoroastrianism, Judaism, Christianity, and Islam...assert the existence of a supreme being...But condition of human society does not remain unchanged and through growth of knowledge, notably in natural sciences, host of causal relations in nature are illuminated. Therefore, religion undergoes reform in step with advancing knowledge. But reformed religion is the beginning of the end of religion.[23]

The new age of industry gives rise to social calamities. Unemployment, crises, bloodletting wars erupt over commercial interests. This creates a deep sense of despair....Gerhart Hauptmann who at age thirty wrote the inspiring *Weavers* [1892], at seventy is sapped of his spirit...Bergson renounces the utility of reason and logic...Mach and cohorts react to modern physics by pointing a revolver at the law of causality ...Spengler prophecies decline and doom of the West.[24]

This state is marked by an overwhelming sense of crisis and leads Taghi Erani to the finale of his discourse on psychology, articulated in two theorems that may be called—the law of mental gyration, and the law of human action.

The Law of Mental Gyration

If man's actions are piloted by the gyrations of the mind, as history is nothing more than the diary of man's actions through the course of evolution, it is by unraveling the workings of the mind that the secret to comprehension of history will be revealed. This extends to how man's actions may be regulated on a sound and rational basis to secure a sane future. After deciphering the "law of mental gyration" shall come its application to the most pressing problems of mankind—man's attitude toward man, a case of self-destructive disorder, and man's attitude toward nature, the equivalent of sitting on the limb of a bountiful tree, plucking off the fruits with one hand, sawing off the branch with the other.

Just as individuals demonstrate neurosis and bouts of psychosis, including self-mutilation, so does humanity demonstrate mania, hysteria, schizophrenia, and a host of other ailments at a collective scale, in various garbs and guises. But just as coping methods exist for individuals, so can they be brought forth for the species at large. At least this is the motivation behind venturing into psychology as the potential savior of humankind. Deciphering the "law of mental gyration," therefore, became the motivation for Taghi Erani to delve into psychology. In his manner of conceiving the subject, this law can be expressed as a trinity of principles.

We are first and foremost biological beings; as such, we are programmed to seek our survival. We may call this the *Principle of Survival*. There is a second principle, which we may call the *Principle of Parallelism*. The individual is the microcosm of society; the mind of the individual a microcosm of the culture of society; culture nothing but the macrocosm of the minds of the individuals comprising society.

The fact that the mind of the individual is a reflection of the character of culture is due to the complexity of the mind, arising from the physiological complexity of the brain and the nervous system, the product of a long train of evolutionary biology. Thus, implicit in this second principle is the *Principle of Neural Complexity*. That is why our species has been able to secure the position that it has secured in nature. The complexity of its cerebral apparatus is such that to make it capable of capturing the complexity of the external world, which exists independent of us and has long preceded us, as reflected in the fossil record.

So implicit in this second principle is the *Principle of Objective Reality of the External World and its Causal Character*. The external world is anything but arbitrary in its makeup and behavior. If its behavior were random, we would never be able to perceive it as possessing predictable behavior.

And if it is argued that it is the mind that fabricates order out of the natural chaos of nature—as some have interpreted Kant in that sense[25]—then how to explain why nature has made such an exception. If nature behaves randomly, then why not when it comes to constructing the brain, which it has mysteriously endowed with the capacity to create order out of disorder.

And more importantly, how to explain our success as a species in reining in on the chaotic forces of nature—the success has been so resounding that it has led to the development of civilization to an extent that state-subsidized armchairs are provided for philosophers to doubt the validity of the existence of the external world independent of man and its causal character. The very questioning of this law is proof of its validity.

The workings of the external world are not arbitrary, and we can be convinced of this based on what we know already about the *modus operandi* of nature, irrespective of our lack of knowledge about many other aspects of it. Lack of knowledge about the causal character of phenomena does not constitute the absence of causality.

As nature wiggles, we are able to register its beat with a high degree of fidelity arising from our capacity to swing from immediate inferences to far-fetched abstractions, as manifested in language; and to archive this information for contextualization of new experiences, a process called learning, as manifested in memory recall and recognition. This attribute of registering the subtleties of nature underpins our ability as a species in scaling the heights of the evolutionary ladder.

The end result of the second principle is that just as individuals can experience neurosis, or worse psychosis, so can cultures and civilizations. Recognition of this parallelism is fundamental to making headway with psychology as the discipline with the potential to provide clues about human behavior: clues with a direct bearing on our species' chances of survival.

In a 1934 essay, *Mysticism and Materialism*, he frames this concept of parallelism as follows:

> Biological function of desire is for the individual to be drawn to things that are beneficial for its survival. Affection is an accentuated form of desire and is also biologically useful. Love is an intense form of desire which blurs the mental faculties and is a passing form of mania. Obsession is an exaggerated form of love, persistent mania, where contents of the mind turn to delusion and fantasy. Obsessive introversion (mysticism) is a social neurosis. It erupts during periods of calamity and although it has a therapeutic value, sedative from a biological perspective, nevertheless it has to be considered a psychic abnormality at societal scale.[26]

In similar vein, Freud had this to say in *Civilization and Its Discontents* (1930):

> If the evolution of civilization has such a far-reaching similarity with the development of an individual, and if the same methods are employed in both, would not the diagnosis be justified that many systems of civilization—or epochs of it—possibly even the whole of humanity—have become 'neurotic' under the pressure of the civilizing trends?....And with regard to any therapeutic application of our knowledge, what would be the use of

the most acute analysis of social neuroses, since no one possesses power to
compel the community to adopt the therapy? In spite of all these difficulties, we may expect that one day someone will venture upon this research
into the pathology of civilized communities.[27]

As is well known, many experiments have been conducted in group psychology, especially after WWII, suggesting a variety of pathological behaviors that may arise in group contexts. Here, however, Freud and Taghi
Erani are referring to a wider sphere of manifestation, that of human society, encompassing human civilization. This suggests that there must be
yet another principle at work; and for Taghi Erani indeed there is; it may
be called the *Principle of Mental Maturation in Human Space-Time*.

This designation may sound like a physical scientist trying to extrap-
olate a concept beyond its domain of application. But in reality, it
only stems from the field of folk psychology, to which Taghi Erani was
attached, following in the footsteps of Wundt, so this principle could
also be called the *Principle of Geographical-Historical Development and
Maturation of Human Mentality*.

For primitive man, there was neither geography nor history in a conceptual sense. He had little memory of his ancestors and little need to
roam beyond his habitat, though it may have been a vast territory. The
culture of primitive man is revealed to us from Paleolithic anthropology. His mentality can be inferred from what we learn in ethology and
primatology. Today we do not know much more than Köhler did a
century ago; perhaps Konrad Lorenz adds a bit more.[28]

For totemic man, there was history but as yet no geography, hence
ancestor worship and primitive cosmogony conceiving the tribe to have
descended from inanimate objects (rocks, meteors), or vegetation and
animals, or all in succession, such as in *Aboriginal Dreamtime*. He is
separated from nature and sees himself in relation to it, as opposed to
belonging to it, just as an infant gradually understands the separation
between itself and the mother. It is a sequence unfolding in time.

But totemic man is not only separated from nature, he is also separated from his fellow man. The distinguishing artifact of this period is
the shield, which is of no use in fighting off wild beasts, but essential
for combatting his fellow man. The growth of the primordial population
advanced to such an extent as to cause perpetual competition and conflict among its ranks. A sense of belonging to herd or group, enhancing

man's chances of survival became uppermost in the mind of man. And belonging to a group, above all, means excluding others from it. We have the coexistence of internally bonded, externally disjointed social units, who may develop symbiotic relationships of exogamy, exchange and barter, but as a rule, they would not hesitate to eliminate each other if all cannot be sustained and accommodated by the environment, which by default is local.

That is how on top of clan and ethnic history, during the next phase of development of our species, geography becomes a reality for man, that is, man of the heroic age; the age about which philology gives us faint notions, the age of onslaught and conquest—*Rigveda, Avesta,* Homer, the Old Testament—reinforced by archeological ruins and tablets. Coexistence having become impossible (crisis of the totemic society), there came migrations, invasions, subjugations, settlements, and new rules of coexistence, including sanctification of enslavement, castes, class, rank, heredity rights, and the like. This presupposes a certain development in productive forces of labor, such that enslavement would be rendered possible (slave labor having to yield more than what is required to sustain the slave, for slavery to proliferate).

From this process of geographical expansion and human conflict, grew the world empires and with it the world religions—Buddhism, Christianity, Islam, the latter two bursting out of Judaism and other thought formations—crystallizing in the mind of man the notion of humanity, not as an emotive outpouring but an objective reality: a totally new concept foreign to all preceding stages of development, for which there was only delimited human material, horde and clan, delimited geographical expanse, gathering and hunting ground, and delimited historical extension, ancestor worship, and tribal cosmogony.

From this point on, human space-time having been exhausted, there is only one path forward—transformation of the human material, or its extinction before the call of nature, reverting back to the inanimate nature from which it germinated. And implicit in this third principle, the *Principle of Mental Maturation in Human Space-Time,* is the *Principle of Maturation-by-Negation.* New thought content comes into existence by negating existing thoughts, but it carries with it vestige of the old thoughts. This is a continual process, the continuity never broken, no matter how radical the change. Disjointed evolution is a logical impossibility; hence, the vestige of the past is always a fellow traveler.

The scientific revolution that five centuries ago set in motion the unleashing of ideas which is uppermost in our minds today is ample proof of the concept of maturation-by-negation. During the medieval period, theology was already being adulterated by science. That is the essence of scholasticism, theology imbued with logic, leading to rational philosophy, in turn giving birth to science, especially when observation and experimentation began to blossom.

This unbroken chain can be traced back to mythology and magic, all the way to the mind of primitive man and the realm of ethology. Magic was a reaction to an experienced reality (for as yet we cannot speak of "observation of reality" which implies the separation of man from nature), as magic at its core rests on an explanatory hypothesis that certain invisible beings exuding with wrath, because of human misdeeds, unleashed their fury, hence the need to appease them. Irrespective of the diversity of such thoughts revealed to us by anthropological fieldwork, at the core of any magic ritual is a hypothesis, and hypothesis is the mother of science. Therefore, magic already contains the germs of what today we call science. And today's science and social norms, when viewed from the perspective of posterity, will no doubt be treated as magic; we need only glance at our present-day conceptions of law, justice, family, inheritance, morals, race, gender, fraught with biases and essentialist categories, to be convinced of that.

In short, the law of mental gyration is deconstructed based on principles of survival, parallelism, and mental maturation, the latter recognizing an unbroken evolution embracing primitive man, totemic man, and heroic man, demonstrating that notions of history, geography, and humanity conceived by modern man are themselves products of historical development.

THE LAW OF HUMAN ACTION

This law is the less complex part of the theoretical construct developed by Taghi Erani in *Psychology* and few key essays—*Man and Materialism*, *Materialism and Dialectics*, and *Materialism and Mysticism*. It too can be elaborated as a trinity of principles—adaptation, population, and contagion.

The first asserts that without adaptation there is only the certitude of extinction. This is not some moral edict, but the verdict of biology.

Thus, if the present conduct of humanity is heading toward the precipice, extinction can be averted only if the mentality of man, acting presently against its self-interests, adapts to its reality. That is, it eliminates the chasm, overcoming the gulf between its conceptions and the objective reality of its conditions, societal and environmental.

The second principle asserts that the reason there is such a chasm is because the human race, like all biological species, consists of population groups; and in population groups there is a spectrum of motives and aptitudes. In fact, this is the only reason why species are not essentialist entities, which if it were so, it would preclude evolution. Rather the natural variability within the population gives it the necessary tension for evolution and progression.

The advantage of sexual reproduction is the constant generation of variability within the population, not so much by mutation as by genetic recombination. The practical consequence of this principle is that in any population, it is the most repressed who are the most motivated to change the status quo; unless, they suffer from neurosis and have resigned themselves to their condition. Once they break this spell, they become vectors of change and hurl the population toward new experiences; hence, force its adaptation and evolution. Here, we have echoes of Rudolf von Jhering,[29] whose conception of natural law Taghi Erani adhered to,[30] the inevitability of strife to effectuate change and bring about social progress.

Finally, the third principle asserts that once thoughts that are impregnated with reality and correspond to sentiments of society, or collective psychology, are articulated, they will gain ground and grow in keeping with the phenomenon of transference and contagion. In this too, there is nothing metaphysical. It is a psychological phenomenon characteristic of organisms endowed with consciousness, social organization, particularly language, driven by instincts of survival and perpetuation of the species.

Taghi Erani, therefore, embraces the concept of human action as free will, and free will as volition, anchored in social and historical conditions. Free will disconnected from concreteness of life is for him a vacuous notion. Reality delineates the scope of our actions, and our actions shape reality. All social organisms evolve in this way, without a predetermined path or a preordained purpose, without adhering to some abstract notion of free will:

Act of volition consists of four stages – cognition, arousal, resolution, and execution; therefore, it results from the [neuro-physiological] structure of the organism and the characteristics of the environment. It is possible to influence the conduct of members of society and to influence and shape the environment...As such, abstract free will is an empty notion, as is pre-destination dictated by divine force...The goal of education is to create the reflex and the judgement to will that which is beneficial and the resolve to execute that which is willed. This holds for individuals as it does for society at large.[31]

Toward Political Activism

To conclude the preceding line of thought, it must be added that just as the law of mental gyration is global in nature—the world is so tightly knit that the fundamental unit of thinking cannot be anything but humanity as a whole (and this has been the case since the age of empires, nowadays dubbed as "global village")—in contrast, the law of human action is only local in nature. One can only act locally, but if the thinking underpinning the action is imbued with a sense of reality at a global scale, it can grow beyond its local confines.

This explains why the works of Taghi Erani, by and large, are in Persian; to his detriment it may be argued, for he did not gain entry into the theater of interwar intellectual history, that is European history. Overall, he seems to have played a high-stakes game of all or nothing— he would leave his imprint on Persia, or wither without a trace.

Besides the inherent riskiness of this all-or-nothing approach to pos-terity, in Persia of the time (witnessing the founding of a new dynasty by a king who ruled with an iron fist), this strategy had another element of risk as well. The course of human history is more significantly impacted by what happens in the leading nations, or the center of gravity of mod-ern civilization given its cultural and commercial grip on the globe and the colossal destructive forces concentrated there. The fate of those oper-ating in the periphery is to a great extent determined by the actions of those operating in the center.

Thus, he must have had the conviction that there are sufficient resources working on that front, or that his addition would be of mar-ginal benefit. Each patch of geography has its own dynamics, and just as things can roll from the center to the periphery, so they can the other way. One has to choose where one wishes to play. And for him, his play-ing field would be Persia.

In addition, he had begun to make his diagnosis of the neurotic condition that in his view incapacitated that culture and social milieu—the masses, especially the youth, taking refuge in the soothing venues of occult, mystical, and gnostic orders, not far from which was the opium den, while the elite aspired to revive the grandeur of the Persian Empire with European coloring, and a middle class seduced by Europeanism, yet attached to traditional values; all while traditionalists that retrenched in their theological obscurantism were tightening their grip on the minds of the rural population and the downtrodden urbanites.

Throwing a monkey wrench in the middle of this mix was perhaps the best way to precipitate its collapse, and liberate its pent-up energies for rebirth and rejuvenation; renaissance pushing forth material and social progress, powered by human energy, science and technology, and nothing else.

Thus, he became the self-appointed psychoanalyst of Persian society, engaging not only in diagnosis but also in treatment planning and execution; psychoanalyst and therapist merged into one. Freud had made the prophecy that psychoanalysts of cultures and societies may one day emerge. In the words of Homer—*as the word was spoken, the deed was done.*[32]

Several passages in *Psychology* give us an eerie feel that as he was working through the academic aspect of the subject, there was a great deal of introspection going on along these lines:

> Changes that occur in the self, take place in both gradual and drastic ways. The individual may perceive a certain social self, but since one's values evolve, the notion of the social self evolves...In youth, the mind may produce an image of a spiritual self, and later scientific self, national self, world self, and so on. In addition to the past self and the present self, there is also the future self. Individuals prone to optimism envisage this as an ideal self.[33]
>
> Sacrifice is a complex sentiment formed principally from social instincts. It requires sound upbringing, suppression of personal interests, and firm logic. It is possible for an individual to conceive a firm relationship between the self and society, undertake deeds that on surface seem to benefit only society. This urge may be accentuated because of certain factors of the environment, subordinating entirely the interests of the individual... Sentiments of sincerity, righteousness, purity, abhorrence of falsehood, are all socially conditioned.[34]

We sense a certain resolve building up for substantive action in the arena of society. Of course, we have the benefit of hindsight. With such benefit, we find even more clues in his works; this time with reference to Rudolf von Jhering, the nineteenth century German jurist, his chief doctrine elaborated in his best-selling work, *The Struggle for Law* (1872):

> The end of the law is peace. The means to that end is war...All the law in the world has been obtained by strife. Every principle of law which obtains had first to be wrung by force from those who denied it...The law is not mere theory but living force.[35]

And this is what Taghi Erani had to say about Rudolf von Jhering:

> On the basis of historical research, one of the notable jurists of Germany, von Jhering, has proven that there is no precept in ethics or jurisprudence that does not flow from stark recognition of societal well-being.[36]

It is to societal well-being, as he saw it, that he turned his energies and attention, and this is how he slid into the realm of political thought and political action with stark consequences.

NOTES

1. He earned 47.25 Marks per month; see *Bankverbindung*, p. 149 (recto–verso).
2. He received 60 Tumans per month for 24 months; equivalent of 300 Marks/month; see Boroujerdi, *Erani Faratar az Marx*, pp. 507, 514.
3. See Erani, *Donya*, 12-issue Volume, p. 251.
4. See Erani, *Osool'e Elm'e Rooh*, p. 180.
5. See Erani, *Donya*, 12-issue Volume, distillation of pp. 68–75.
6. See Erani, *Osool'e Elm'e Rooh*, p. 159.
7. See Erani, *Donya*, 12-issue Volume, p. 334.
8. See Erani, *Osool'e Elm'e Rooh*, p. 13.
9. See Erani, *Osool'e Elm'e Rooh*, p. i, of manuscript at McGill University (Institute of Islamic Studies Library) starts with the sentence given in the text—"Psychology forms the fourth book in the Series in Exact Sciences." Later reprints of *Psychology* have modified this first sentence as: "Psychology is one of the subjects of Series in Exact Sciences" (1978 reprint, Tehran, p. 5). I have given the original version in the text.

10. See Erani, *Osool'e Elm'e Rooh*, p. 14.
11. Ibid., p. 69.
12. Ibid., p. 36.
13. Ibid., p. 52.
14. Ibid., pp. 10–11.
15. Ibid., p. 56.
16. Ibid., p. 57.
17. Ibid., pp. 79–80.
18. Ibid., p. 62.
19. Ibid., pp. 241, 250.
20. Ibid., p. 9.
21. Ibid.
22. See Erani, *Donya*, 12-issue Volume, p. 176.
23. See Erani, *Osool'e Elm'e Rooh*, p. 271.
24. See Erani, *Donya*, 12-issue Volume (in the order quoted), pp. 126, 119, 120, 268, 120.
25. Ibid., p. 118.
26. Ibid., pp. 321–322.
27. See Freud, *Civilization and Its Discontents*, pp. 141–142.
28. See Lorenz, *Behind the Mirror*, pp. 113–166.
29. See von Jhering, *The Struggle for Law*, pp. 1–2.
30. See Erani, *Donya*, 12-issue Volume, p. 150.
31. See Erani, *Osool'e Elm'e Rooh*, pp. 253–257.
32. *Iliad* (Book XIX): "Forthwith as the word was spoken so was the deed done."
33. See Erani, *Osool'e Elm'e Rooh*, p. 141.
34. See Erani, *Donya*, 12-issue Volume, pp. 222–223.
35. See von Jhering, *The Struggle for Law*, pp. 1–2.
36. See Erani, *Donya*, 12-issue Volume, p. 150.

CHAPTER 10

Debut in Political Activism

Taghi Erani's political activities began in Berlin. The gap between overt and covert action was slight initially, but with time it grew and reached daunting proportions with fatal consequences. What mind-set and circumstances brought this about and how it unfolded from Berlin to Tehran is the real story of his life, and all that we have seen thus far is only a prelude to the real action, the climax of his life story.

There is a saying in Persian that there is nothing but somberness and obscurity when it comes to the origins of politics, its paternal and maternal lineage—*siasat pedar va madar nadarad!*—that it is a bastardly child, a savage devouring all that flirts with it, from kings and ministers, to lords and chieftains, prophets and priests, radicals and revolutionaries, poets and romantics, heroes and heroines, the whole lot.

Persia's history bears this out like an immutable law of nature, yet others come forth and the pathos and drama repeats itself with startling regularity. What is it that gives this propensity to suffering and sacrifice to this people is baffling. The life story of its tragic figures gives us a glimpse into the mental universe gripping the people of this land.

Taghi Erani left Berlin in a rush, and that was perhaps his most judicious reflex from the standpoint of survivability. The art of dissimulation is second nature to Persians in light of their history—constant onslaught, punctuated by interludes of physical and mental reconstruction, a history stemming from its peculiar geography and demography, in western Asia,

© The Author(s) 2019
Y. Jalali, *Taghi Erani, a Polymath in Interwar Berlin*,
https://doi.org/10.1007/978-3-319-97837-6_10

at the crossroads of many peoples. Reconciling continuity of one's traditions with the need for conformity to abrupt external realities cultivated a culture of dissimulation, as protection against the vagrancies of life.

Taghi Erani, perhaps, concluded that if he refrained from returning, that if he became a dissident in exile, he stood no chance of imparting influence on the course of events in Persia. He likely deliberated that his force would come only from bonding with the soil and fabric of Persian society, that deprived of his homeland, he would drift into irrelevance and obscurity. This theme is evident even today in the trials and tribulations of Iranian artists, intellectuals, and political actors.

For Taghi Erani, it amounted to a split-second decision. If he stayed behind, the prospects were abysmal, socially, culturally, and politically. The best course of action was to return home. Returning was itself proof of no wrongdoing. He may have had a slip here or there, with the excesses and exuberance of youth, but nothing outrageous. If a reprimand were to come, the most likely outcome would be a slap on the back of the hand, or something slightly more severe, but then he would be good to go for the next leg of the expedition, albeit in treacherous territory. But for that subsequent phase, other stratagems could be conceived in due time. For now, it was imperative to overcome the predicament staring him in the face. He had been summoned by the government to return home without delay.[1] Given the leverage of the government of Persia on the German Foreign Office, and their leverage on the Ministry of Interior, it was not unlikely that they would force his expulsion, if he chose not to comply, though he could still play the Rosen card or go to Prague or Vienna and buy time while pondering his next move.

There is a good deal of conjecture in all this, not in terms of external circumstances but in terms of his state of mind. However, when we put together all the pieces, it is difficult to reach a radically different conclusion in terms of his mental calculus. Leaving was the best move, although he had to part with much that he found exciting in Berlin.

We know little about his personal life, only that in the police interrogations he indicated one reason for his return to Berlin in the summer of 1935, was the possibility of marriage.[2] Family archives have several photographs of Taghi Erani from 1920s and his two trips in 1930s, in the company of female friends, the same set of faces are seen in all the photographs, notwithstanding the six-year gap between the date of departure and his first return (December 1928, September 1934).

Irrespective of the companionship question, we know he was impressed by the strides women had made in Europe, engaged in cutting-edge science and technology, shattering the gender chasm, in stark contrast to Persia. This, despite the fact that he judged the condition of women, even in the West, to be one of persistent and structural subjugation.[3]

We also know of his love of opera, and the soft spot he had for the tragic story of Cho Cho San in *Madame Butterfly* that he had seen in Berlin.[4] He was not a practitioner of performing arts, and admitted as having no particular skills in music, but he was drawn to it intellectually. There are remarks in his writings about classical music and symphonic performance; thus, it is not unlikely that at *Kroll* or *Linderoper* he attended not only operatic but also orchestral performances.

His own voice is described as having had a metallic resonance,[5] and by the admission of many he was a captivating orator; it is a conjecture he was character grade tenor, Francesco Tamagno a likely analogue.

He has numerous allusions to western poetry[6]—Schiller especially, and Lamartine (*Le Lac*)—and he found a certain poetic excitement in the frenzy and commotion of industrial production:

> In this exhilarating world of steel and steam...when science leaves the tranquility and tidiness of the laboratory for the fervor and fury of the factory...[7]

In the final months of his stay, he was engaged in work of an industrial nature. It is not clear how this connected with his doctoral work, but shortly before departure he jotted these words on a postcard to his mother:

> My Dear Mother; your devoted; I am busy with operations in a factory, and soon will write at length in response to the handwritten note of your pureness; know you will be delighted of its contents; for now, I kiss your hand, your devoted son, Taghi.[8]

He was also engaged in his teaching. The winter semester had begun in October and would run until the following March. But in December he had to break it off, and in a hastily drafted petition to Prussia's Minister of Culture, request a six-week leave of absence.[9] As archival records reveal, Rosen was in the know and urged the institute to give their consent.[10] As it turned out, he would leave and not come back for another six years.

Finally, Erani had to pull the plug on his political activities, which simmered on both the front and the back burner, overt and covert action. It was only in November 1928 that he pleaded to Rudolf Breitscheid for his backing to prevent the expulsion of a fellow student engaged in anti-Pahlavi political activism.[11] And it was in December that Breitscheid's appeal came out in *Vorwärts* (circulation figure half a million, December 19, 1928 issue, *Die Republic lässt lich missbrauchen, Liebesdienste für den Schah—Republic Abused, Love Affair with the Shah*).[12] There was, however, a lot more to the political side than these outcrops would indicate.

***

It all began in a very gentle and benign, almost fraternal way. There were six expatriate journals in 1920s Berlin put out by the Persian diaspora, a community of one hundred and twenty. One of these had ceased publication before Erani's arrival, the other five coinciding with his stay, and in all five of these there are contributions from Taghi Erani.[13] Not that he was a prolific contributor or even the editor of any of these; his contributions are ten in number as far as we can determine, one to four contributions in each of these journalistic publications. The thrust of which was cultural, with political undertones, explicit occasionally.

The reach of these journals while limited was still noteworthy. It circulated in other European cities, as well in Constantinople, and occasionally also back home in Persia. But that was the extent of it, and some are extremely short-lived, as short as several issues in as many months; the longest lasting four years.[14]

During the war, most if not all hopes were pinned on a German victory. In fact, when the Eastern Front collapsed, there was a big celebration in Berlin by the Persian expatriate community;[15] the Russians had been routed! Through the channels available to them, the leading figures of the diaspora sent off details of what could be included in the Brest-Litovsk agreement, as far as Persia and the Caucasus were concerned. By some accounts, the Germans did take these into consideration.[16] One can easily understand the cause of the jubilation; here was Persia's lifelong nemesis reduced to defeat. The nation that had most molested Persia and hurt its sense of pride and self-esteem had finally gotten a bitter taste of its own medicine. Poetic justice! What could be more sublime than that?

Colonel Pasyan, a daredevil who volunteered on the Eastern Front as a pilot and had won accolades (the Iron Cross First Class), was the guest of honor on this occasion,[17] but then Germany imploded. The sustained naval blockade took its toll, the turnip winter, the collapse of Germany's principal ally (Austria-Hungary), the desertion and mutiny (Kiel), the fleeing of the Kaiser (the Netherlands), and the collapse of the empire. Thus, the chilling realization that if the Persians wanted liberation, to shed the shackles of Anglo-Russian usurpation of their country, they had to do it on their own. And possibly the deeper realization that even if the Germans had won, it would not have altered the fate of Persia. There goes one, in comes another; a change of taskmaster, guard, warden.

This, plus the turmoil in Russia, stirred by the revolution and civil war, which meant that they were out of action in Persia, and predatory moves by Great Britain, with which the corrupt politicians in the capital colluded, led to a wave of nationalistic movements, especially in the northern provinces. The Berlin crowd who were personal acquaintances of leaders of these insurgencies watched the turn of events with throbbing hearts.[18] When all those fell like dominoes, their leadership decapitated—some literally—emotions hit rock bottom. The worthies of Berlin threw in the towel; their nerves could no longer take it.

One became a Bonapartist (enlisting with Reza Khan, the chief saber-rattler, who pulled the strings in the slaying of the insurgents, handpicked by General Ironside for this role, the British wanting not only firm control over Persia but also a dyke against Bolshevik spillover). Another became a chronicler of Persia's modern history with a predilection for ethics and morality. Later he veered towards theosophy. A third ran for cover in the soothing refuge of fiction and literature. A fourth sought solace in the edifying rituals of pure orientalism. Another sank into depression and committed suicide on Berlin train tracks a few years later. Such were the reactions of Taghizadeh, Kazemzadeh, Jamalzadeh, Qazvini, and Alavi-Senior. The rest simply scattered. This was the older generation (born late 1870s to early 1890s).

When the next batch came in, the fervor erupted once again. When the dynastic upheaval was fully consummated (the fall of Qajar, birth of Pahlavi, December 1925), some saw it as the eternal salvation; here came Persia's Napoleon, Reza Shah Pahlavi, rescuing the ancestral land from the abyss and chaos of history; whereas a few others, Taghi Erani included, saw it as a tumor, sapping Persian society and culture of its life and vitality, surrendering it to the dark forces of colonialism, feudalism, militarism, and chauvinism, for the sordid interests of a wretched minority.

There could be no greater gulf between these two camps and conceptions. Precisely for this reason, the chance encounters and coalitions that had come into being in the journalistic arena sublimated into thin air. Hence, the appearance of a second phase in the political life of Taghi Erani in Berlin, one marked by pugnacity and covert action. On one track, above ground, the semblance of continuity with the past, journalistic collaboration in the cultural realm with political undertones; and on the other, a hardened political rhetoric and plan of action. The latter came with a certain fervor akin to religious conversion.

Taghi Erani deserted a good part of his earlier views, in particular, on the mythologizing of ancient Persia. He began to view history and society in radically different terms. His studies in orientalism, rooted in ancient and medieval history, and his geographical broadening beyond Persia, as well as Persia's contemporary events, along with his reading of modern political theory and history, all reflected in his writings, made it plain to him the naivety and superficiality of his earlier perceptions. He may have realized that his beliefs stemmed from a psychological predilection for self-delusion; emotions and urgings, not facts and logic. Whatever it was, the result was a dramatic shift without reversion, altering his life entirely. This is what set him apart from his cohorts; at any rate, those who outlived him by a large margin.

***

Of the ten journalistic pieces done by Taghi Erani in Berlin, two came out in a journal called the *Liberation of the East* (*Azadiye Shargh*). This was a loaded name, as it begged the question—liberation from whom and by whom. And the message the editor sought to pass, rather subliminally, was—liberation from the claws of Anglo-Russian hegemony, and by who else than the Germans. This was a theme that had been concocted during the war, and it showed surprising resiliency even after the war. The same editor put out another journal called *German Industry and the East* (*Sanaye'e Alman va Shargh*), and again the motive was soft propaganda. The editor, Abdol-Rahman Seif'e Azad, was a shifty character and had links to German political circles associated with the security apparatus.[19] We do not know much about him, other than the fact that he put out a mouthpiece for pan-Aryanism during the reign of Reza Shah, published in Tehran (*Iran'e Bastan*, Ancient Iran), backed by Hitler's propaganda chief, Goebbels.[20] We also know of someone who followed in his footsteps in the mid-1930s (M.-H. Ayrom),[21] posing as someone who could shepherd the Germans into Persia by virtue of his insider knowledge. Taghi Erani published two patriotic poems in the first journal, titled *Motherland* and *Mohammad*, and a technical piece in the second, marveling at German strides in railroad construction.[22]

Then came four articles in another journal (*Iranschähr*, connoting land of Iran), put out by a gentleman steeped in paleo-nationalism (inspired by mythical conception of antiquity), and drifting towards ethics, morality, and theosophy (Kazemzadeh). He was the polar opposite of the previous character; one unscrupulous, the other a self-cultivated moral philosopher. Two of the pieces in this journal written by Taghi Erani were extremely short;[23] one was a questionnaire put to the readership, asking them to name five of the greats of Persia's history, in two categories—cultural and political. The second was a tally of the winners of this contest, alluding to Persia's grandeur and glory of the past. The third was a piece on the interplay of culture and technical savviness,[24] and the fourth, a piece on the linguistic deficiencies of modern Persian.[25] The latter marked his debut in orientalism. This article too had political undertones depicting the Russification (on the flanks of the Caspian Sea) trampling on the Persian-speaking world and extinguishing the fountainheads of its classical literature.

Then came his most virulent piece, in yet another journal (*Nameye Farangestan, Letter from Europe*, editors, M. Kazemi and A. Farhad), put out by the association of students from Persia, residing in Germany, group called *Aspiration of Iran* (*Omid'e Iran*), also referred to as *Iran Society* (*Majma'e Iran*). This piece was called *Azerbaijan, a Matter of Life and Death for Iran*.[26] It bashed at Ottoman historiography which attempted to prove that this province was an integral part of the ancestral land of the Turks. Taghi Erani did not react to this claim well, noting that the Ottoman Empire was only a few centuries old, and statehood amongst Turks a bit longer, if we account for the Seljuqs of Anatolia (Minor Seljuqs). On the other hand, Azerbaijan has several millennia in recorded history, with Zoroastrian temples, and it was the theater of war between the Persians and the Romans, and later the Persians and the Arabs, before the Turks entered the arena of history. He went so far as to state that the people of that territory are not only proud of their Iranian heritage, but also that they are ashamed of being identified with the Turkish culture and its language. This last assertion was one too many, as he admitted years later,[27] but he said what he thought, and he said it as a native Azeri. His own family history was a fallout of Russian expansionism preying on this region; and here was another, a remnant of the Great War and Ottoman designs on this territory, coming out in the guise of revisionist historiography.

Finally came two pieces in yet another journal (*Elm va Honar, Science and Art*) edited by the gentleman who aspired to become a novelist (Jamalzadeh). The timing of these publications (1927) coincided with when Taghi Erani's covert activities were in full swing. It served partly as camouflage and partly as the leveraging of available outlets for his political aims, in the guise of commentary on science and technology. One of these articles narrated the progress in science and the arts in Europe since the nineteenth century, stressing its tempo since that datum.[28] The other reiterated the same theme, contrasting the dynamic landscape of Europe with the stagnation of Persia,[29] the persistence of superstition and occult schools of thought distrustful of science, as if the civilizational lacuna could be overcome by chants and meditation.

> The foremost thinking of humankind flows from the natural sciences, and its fundamental facilitator is the logical mind...Without logical and mathematical thinking, man cannot undertake investigations in any science. The three sciences – physics, chemistry, and physico-chemistry (which in addition to biology are called exact sciences) – rely on logic and mathematics more than the other sciences. In Persia, however, there is a group of people who are staunchly opposed to mathematics, that is, they do not subscribe to rational thinking [rather to revelation and illumination]. These contrarians are a curiosity in the world of the twentieth century.[30]

***

Iran Society was not set up as a front organization. It came about spontaneously, as an increasing number of students from Persia arrived in Berlin, and they set up their cultural newsletter (*Nameye Farangestan*).[31] When it became evident that the Qajar was destined for the dustbin of history (Ahmad Shah was forced out of Persia in 1923, presumably for medical treatment, settling in Paris, and never returned), and Reza Khan assumed increasing authority (Head of Army, Minister of War,

Prime Minister, accumulating all posts), tensions accentuated amongst the diaspora. From amongst the membership of *Iran Society*, a covert group emerged, possibly a dozen in number, who viewed these developments as disastrous for Persia.[32] The usurpation of power by a Cossack clique was trampling on the constitutional rights of the people and plundering the country.

However, there was not unanimity of purpose amongst this covert group. Some insisted on the respect for the constitutional revolution and what it had enshrined (constitutional monarchy, 1906). Others went further and wanted to do away with the monarchy, advocating a republic anchored in social justice and rehabilitation (as opposed to the republic advocated by Reza Khan as a tactical ploy for accession to power, exposed at the time by the poet Eshghi, who was assassinated at his home).

When Reza Khan took the last step, coerced the *Majles* to vote the abolition of Qajar and founding of Pahlavi, with himself as the absolute monarch, and exclusivity of succession in his bloodline, the covert group fell apart, and from it was born a hardcore group with revolutionary zeal, operating in stealth.

This group adopted the name CRRP—*Circle of Revolutionary Republic of Persia* (*Fergheye Jomhooriye Enghelabiye Iran*)—with an elliptical seal with which all its correspondences were stamped.[33] Based on documents from four sources—Persia's Ministry of Interior (memoranda declaring discovery of CRRP literature intercepted in the postal service), Russian archives (Comintern), German Foreign Office, and the British Foreign Office—we now know that this group consisted of five individuals, who identified themselves as the "Central Committee" of the CRRP, suggesting they also had members or sympathizers. The five principals were all members of *Iran Society*. With the exception of the Comintern, that knew the identity of all five, all other services knew the identity of only two, and this with a separation of two years.[34] They did suspect a third member, but their investigation erroneously indicated that the suspicion was unfounded.[35] At any rate, none suspected Taghi Erani to be a member of this group. He was the secretary of the *Iran Society* (succeeding R. Schafagh), and this gave him the legitimacy to protect the rights of its members, including prevention of attempts at expulsion. His credentials made him a natural candidate for this position—scientific work, orientalist work, and cultural journalism.

The main political actions of the CRRP included—protest against the regime of the Cossacks and Reza Shah, coinciding with his coronation (April 1926). This was a gathering in Berlin to which at least one prominent member of German left was invited (Georg Ledebour, USPD); it was a clandestine gathering.[36] Participation in the anti-colonial conference held in Brussels (February 1927); agitation against Persia's Court Minister, Teymurtash, on a visit to Europe (September 1928); and political publications (1926–1928, ten pamphlets, exceeding one hundred pages, in Persian with French translations, with mailings to Persia, India, Turkey, and major European cities, some of which were intercepted, as reflected in government archives).

They also undertook initiatives for collaboration with other organizations. This included submission of their political charter to the Eastern Bureau of Comintern (via Louis Gibarti, liaison in Berlin, a Hungarian whose real name was Laszlo Dobos, a mysterious character who collaborated with US intelligence post-WWII); contact with the International Worker's Relief (IAH, W. Münzenberg); participation in the anti-colonial conference in Brussels organized by the League Against Imperialism (LAI, W. Münzenberg); collaboration with the Socialist Party of Persia (SPP, S. M. Eskandari, *Majles* deputy, party outlawed by Reza Shah) in the form of joint resolution submitted to the Brussels conference; and solicitation of support from the Independent Social Democratic Party of Germany (USPD, G. Ledebour, R. Breitscheid).

***

The five founding members of CRRP were—Ahmad Asadi, Taghi Erani, Mahmoud Pour-Reza, Ahmad Farhad, and Morteza Alavi.

Asadi acted as the front for the group. He was a student of political science at the University of Berlin. He was the speaker at the closed-door event (April 1926) exposing the ascension of Reza Shah to power. He used the codename "Darab" and had connections with the German Communist Party (KPD, Carl Wehner).[37] He is the one who was identified early on by the German authorities. When Teymurtash visited Berlin (September 1928), a tract was distributed exposing the Court Minister's background and involvement in corruption in Persia. (He had executed insurgents in Gilan when he was the governor there, had amassed a personal fortune by collecting illicit commissions on behalf of the Shah and himself from concessions granted to foreign concerns, and after the Shah was the most powerful man in the country, overshadowing the Prime Minister.)

This action against Persia's Court Minister in Europe led to a flurry of diplomatic exchanges in which the British government got involved, backing demands by the government of Persia for the expulsion of Asadi, asserting that he is a member of a subversive organization whose literature is inciting anti-British sentiments in India.[38] This issue escalated from the Foreign Ministry and Interior Ministry to the chancellor's cabinet (Hermann Müller);[39] but before this was officially discussed, the German police, under the jurisdiction of the Interior Ministry, expelled Asadi from Germany.[40] This gave the chancellor some measure of deniability. It saved him from criticism in the Reichstag from the left (KPD, USPD) to have given in to coercion from a despotic regime, and allowed him to mend the relationship with Great Britain and Persia, the latter in view of escalating trade. It is believed that Carl Wehner, operating out of Haifa,[41] facilitated the departure of Asadi from Berlin, via Prague and Vienna, and dispatched him to Moscow. Later, he managed to return to Persia and became a Pahlavi dynasty diehard, a turncoat.[42]

The second to be faced with the threat of expulsion was Taghi Erani. As indicated, he was not suspected of being a member of the CRRP (his scientific and scholarly garb gave him a natural cover), although he had acquired a reputation amongst the diaspora as being left-leaning.[43] What brought on the wrath of the government of Persia was his initiative to try to prevent the expulsion of Asadi, by going the extra mile to visit a member of the Reichstag and asking for his intervention to block this move. And worse, that this political celebrity had jumped on the occasion and publicized the affair in *Vorwärts*, the leading daily in the country, with a circulation exceeding half a million.

In doing so, Breitscheid scored a point against Chancellor Hermann Müller, who although from the social democratic party, had formed a coalition with center-right parties, far from the independent social democrats to which Breitscheid belonged. There were major gaps in both domestic and foreign policy between the two, the latter as it concerned Müller's rapprochement with France and Great Britain, and fallout with Russia. Müller also banned May Day demonstrations, which was rather odd for a social democrat. The fact that Breitscheid's article appeared in *Vorwärts*, which was the political organ of the social democratic party, with a spectrum of tendencies, gave his jab an extra punch.

There is a letter from the Persian Minister in Berlin (Azodi) addressed to Fritz Grobba (responsible for Persia, Afghanistan, and India, reporting to Gustav Stresemann, Foreign Minister), asking for the expulsion of Taghi Erani.[44] The letter is dated November 29, 1928. This was after Erani's visit with Breitscheid, but before the *Vorwärts* article (December 19, 1928). Somehow the embassy knew about his movements. It was these events, happening in quick succession, which led to his decision for departure from Berlin. This caused the collapse of the CRRP, although German police records indicate a gathering of sympathizers in Cologne as late as 1931.[45]

Pour-Reza was also a student at the University of Berlin studying law. He is the one the embassy suspected of being a political agitator and was investigated by the German police, but they found no evidence of involvement. (The embassy provided names of three students, all three of whom were cleared by the German police, Pour-Reza incorrectly.[46]) Yet his name is present in the Comintern archives as signatory to CRRP documents.[47] He was also present at the Brussels conference.[48]

Farhad had also been a signatory to the CRRP documents.[49] He was quite active in *Iran Society* and was the co-editor of its newsletter (*Nameye Farangestan*). Sometime after the departure of Taghi Erani from Berlin, he left for Heidelberg and later Frankfurt and became a radiologist.[50] Farhad and Pour-Reza returned to Persia, possibly in mid-1930s, and both stayed aloof of politics during the reign of Reza Shah. Therefore, it is a certainty that even if they did meet with Taghi Erani, they no longer had any political involvements. During the reign of Reza Shah's son, however, there was a showing by Pour-Reza (as one of the attorneys of the plaintiffs against Reza Shah's Chief of Police and his cohorts, 1942–1943). Also, during the reign of the son, Farhad dabbled in politics from the fringes;[51] he became the Chancellor of the University of Tehran but resigned his post in protest to police brutality inflicted on the students, an assault ordered by the Shah and executed by his elite commandos (1962 with fatalities). Thereafter, he dropped out of politics.

Morteza Alavi was a student of economics at the University of Berlin, and signatory to CRRP documents.[52] He is the one whose father had committed suicide, depressed by the repeated failures of Persia's nationalist movements and tragic fall of its leaders. Morteza Alavi stayed in Berlin after the departure of Taghi Erani in December 1928. But in October 1931 he too was expelled. He was identified by the German police sometime in 1930; his girlfriend was presumably an undercover agent.[53] He moved to Prague, then Vienna, Moscow, and Tashkent. He and Taghi Erani managed to keep a certain level of contact,[54] and though each had his own political destiny, the two formed part of a bigger landscape.

NOTES

1. See *Bankverbindung*, p. 103.
2. See Momeni, *Parvandeh*, p. 242, on Erani's allusion to prospects of marriage.
3. Erani's view of status of women in the west, achievements and gaps—see *Bashar az Nazar'e Maadi*, in Maghalat, p. 74; see *Osool'e Elm'e Rooh*, p. 295 on abortion; see *Donya*, issue 6, highlighting women in science; see also *Honar va Materializm* in *Donya*.
4. Erani's interest in opera; see Khame'i, *Panjah Nafar va Se Nafar*, p. 14 (attending Madame Butterfly); see Erani, *Bashar az Nazar'e Maadi*, in Maghalat, p. 81, on Wagner's Nibelungen.
5. See Tabari and Alavi, *Zendegi Nameh'e Dr. Erani*, p. 56.
6. See Erani, *Osool'e Elm'e Rooh*, p. 16 (Schiller), and p. 156 (Lamartine).
7. See Erani, *Donya*, 12-issue Volume, pp. 5–7 (on modern industry), and p. 204 (on Klingenberg power plant).
8. For Erani postcard to his mother, see facsimile in Chaqueri, *Taghi Erani dar Ayeneye Tarikh*, p. 202.
9. See *Bankverbindung*, p. 103.
10. Ibid., p. 134 (recto–verso).
11. See Mahrad, *Die deutsch-persischen Beziehungen von 1918–1933*, pp. 331–337, and Ahmadi *Tarikhche Ferq'e Jomhooriye Enghelabi'e Iran*, pp. 24–25.
12. For Breitscheid's article in *Vorwärts (December 19, 1928)*, see Mahrad, *Die deutsch-persischen Beziehungen von 1918–1933*, p. 335-vero; it has facsimile of this article (*Die Republic lässt lich missbrauchen, Liebesdienste für den Schah*).
13. See Mahrad, *Die deutsch-persischen Beziehungen von 1918–1933*, pp. 476–477.
14. *Iranschähr* lasted four years in two two-year cycles: 1922–1923 and 1926–1927.
15. See Behnam, *Berliniha*, p. 44.
16. Ibid.
17. Ibid.
18. Pasyan was personal friends with Hossein Kazemzadeh Iranschähr. See Afshar, *Zendegi'e Toufani*, p. 361, for Taghizadeh's account of Pasyan correspondence with Berlin expatriates, inviting them to join his government in Khorasan. Ghanizadeh, Berlin based, had participated in the Gilan insurgency and knew Mirza Kuchak Khan. Taghizadeh, leader of the Persia Committee in Berlin, knew Heydar Amoughli, one of the leaders of the Gilan coalition. They all knew of Khiabani from Tabriz connections.

19. Abdol-Rahman Seif'e Azad, journalist; see *Khaterat'e Bozorg Alavi*, p. 105 (compiled by Ahmadi) for linkage of this journalist to a certain von Meyer in German security apparatus. Alavi calls Seif'e Azad "*haft rang*" (seven colors), meaning a shifty character. There is a description of him also in Sadr-Hashemi's work on history of journalism in Persia.
20. See Elwell-Sutton, *Modern Iran*, p. 167.
21. See Amanat, *Mohammad-Hosayn Ayrom*.
22. See Erani, *Qasideh Madar'e Meehan*, *Qasideh dar Molood'e Hazrat'e Rasul*, and *Rahahanhaye Alman*.
23. See Erani, *Soalat'e Elmi*, and *Nam-daran'e Iran*.
24. See Erani, *Teknik, Fann va Farhang*.
25. See Erani, *Tadghighat'e Lesani*.
26. See Erani, *Azerbaijan, Yek Masaleye Hayati va Momati'e Iran*.
27. See Erani, *Taghir'e Zaban'e Farsi*.
28. See Erani, *Tarvij'e Sari'e Elm va Honar az Gharn'e Noozdahom be Baad*.
29. See Erani, *Oloom va Khorafat*.
30. Ibid.
31. Persian diaspora, 1920s Berlin, cultural society; see Behnam, *Berliniha*.
32. Ibid., pp. 61–77.
33. See Ahmadi, *Tarikhche Ferq'e Jomhooriye Enghelabi'e Iran*.
34. Ahmad Asadi and Morteza Alavi, CRRP founders, identified by German police in 1928 and 1930, respectively.
35. See Ahmadi, *Tarikhche Ferq'e Jomhooriye Enghelabi'e Iran*, p. 25.
36. See facsimile in Chaqueri, *Taghi Erani dar Ayeneye Tarikh*, pp. 169–171.
37. See report of German police linking Ahmad Asadi to Carl Wehner reproduced in Boroujerdi, *Erani Faratar az Marx*, pp. 454–455; in Mahrad, *Die deutsch-persischen Beziehungen von 1918–1933*, the spelling is given as Carl Wehner (Wener has also been cited).
38. See Ahmadi, *Tarikhche Ferq'e Jomhooriye Enghelabi'e Iran*, pp. 24–25.
39. Ibid.
40. Ibid.
41. On Wehner's Haifa base, see Boroujerdi, *Erani Faratar az Marx*, pp. 454–455.
42. S. M. A. Jamalzadeh, personal communication.
43. See Momeni, *Parvandeh*, p. 253.
44. See Chaqueri, *Taghi Erani dar Ayeneye Tarikh*, p. 208.
45. See Boroujerdi, *Erani Faratar az Marx*, pp. 454–455 (gathering of February 17, 1931 in Cologne).
46. Ahmad Emami, Mahmoud Pour-Reza, Raazi Eslami were suspected of membership; see Ahmadi, *Tarikhche Ferq'e Jomhooriye Enghelabi'e Iran*, p. 25.

47. See Chaqueri, *Taghi Erani dar Ayeneye Tarikh*, pp. 158, 162, 170.
48. See Petersson, *"We Are Neither Visionaries nor Utopian Dreamers,"* listing Pour-Reza as an attendee in the Brussels Anti-Colonial Conference.
49. See Chaqueri, *Taghi Erani dar Ayeneye Tarikh*, p. 171.
50. See Vessal et al., *Development of Radiology in Iran*.
51. See Harvard Fellows College, *Iran from Religious Dispute to Revolution*, p. 187.
52. See Boroujerdi, *Erani Faratar az Marx*, pp. 450–451, and Chaqueri, *Taghi Erani dar Ayeneye Tarikh*, p. 179.
53. See Ahmadi, *Tarikhche Ferq'e Jomhooriye Enghelabi'e Iran*, pp. 31–32. Also recounted in *Khaterat'e Bozorg Alavi* (compiled by Ahmadi).
54. See Momeni, *Parvandeh*, p. 246, about the codebook sent by Morteza Alavi to Erani; also see Khame'i, *Panjah Nafar va Se Nafar*, p. 71.

CHAPTER 11

Political Reflections

We observe in the works of Taghi Erani an evolution from romanticism, in the sense of emotive attachment to Persia as a fallen nation, to ethno-nationalism, Persia's perceived purity of ethnic heritage, to paleo-nationalism, belief in the grandeur of its antiquity, to a radical departure from this outlook and ideological sea change.

This departure was marked by a broad conception of history as a succession of jolts in man's material and mental culture, bringing about cleavages and schisms within society, with an imprint of tenets of anthropology, cultural psychology, and Marxism.

The deepening of this process of social segregation led to globalism, conflict, conquest, assimilation, and subjugation. This transpired to the present state of the human condition, as manifested by the lopsidedness in the balance of power, leading nations and subject peoples.

This configuration portending towards the future as a contingent process; elemental forces producing an aggregate outcome with unintended and unforeseeable consequences, shattering the notions of determinacy, finality, and teleology.

The outcome of this process conditioned by the collective consciousness and volition of humankind, global culture, ever evolving, in the same vein as natural history, and subject to the same laws—strife, adaptation, inheritance, evolution, transformation, and cataclysm.

In the realm of social theory, this translates to transience of all social structures, the continual process of differentiation and segregation from within (formation of castes, classes, strata), aggregation and coalescence (federation, association, alliances), followed by disaggregation and disintegration and recombination, in constant flux.

The nation-state as the most recent product of this process, with a certain degree of demonstrated stability and longevity. A nation as a spatial and geographical concept, the sum total of human groupings which the historical process has superposed within a geographical realm with intertwined relationships underpinning its shared interests to secure its survival and livelihood.

National aspiration being a temporal concept; nations striving to attain that which modern temporality has demonstrated to be attainable with current knowledge, in material and cultural terms, through mastery of science and technology and social organization.

State and governance being a dynamic concept; tension between social cleavages; national cohesion secured through continual juggling of opposing interests, the dichotomy of right and privilege, preservation of tradition and departure from it, continuity and discontinuity.

This dichotomy being latent in the notion of nationhood; crystallized as the imperative of inventiveness and resourcefulness internally, openness and exchange externally, but in fragile equilibrium. National autonomy is secured through continual juggling of opposites; volatility, unpredictability, its chief characteristic.

In contrast to paleo-nationalism, with which Erani's views were marred in his youth, his mature views may be characterized as nationalism imbued with historicism, social constructivism, globalism and pragmatism; national realism being a possible appellation.

While all this may sound abstract, there was nothing abstract in terms of the circumstances that brought about this ideological transformation. Besides the scientific training he had come to acquire while in Europe, encompassing hard sciences and humanities, there were a number of pivotal events which acted as catalyst to this shift in outlook; events which are chronicled in his writings, notably the editorial column in his journal *Donya*, on "The World Scene"(*Manzare'ye Donya*).

This included the 1929 stock market crash, which happened in the wake of his return to Persia, the havoc in industry and business in Europe, America, and Japan, social and political havoc, particularly in Germany. These jolts led to the emergence of the Third Reich (1933), envisaged by western European powers as the gravedigger of Bolshevism, while increasingly it posed as the gravedigger of liberal capitalism. Then came Great Britain's vacillation between appeasement and confrontation with the Third Reich, the formation of the United Front against fascism, the rapprochement with the Soviet Union, the coalescence of fascism and militarism in Europe and Asia as a political bloc and virulent ideology (Central Powers).

On the home front, what alarmed Erani was the reception of this creed by Persia's ruling elite which was an integral part of a diversification strategy. This involved engagement with Germany on the industrial and macroeconomic front (Schacht), with Italy as potential partner and player (Teymurtash and Mussolini, navy project). This was undertaken while sustaining linkage with British colonial rule, with a naval base in Bushehr and Port Shahpur. The British used petroleum resources, extracted for the consumption of the British navy; the thirty-year concession was renewed for sixty years.

Further, there was the cultivation of a homegrown brand of fascism that was resurrecting notions of Persia's racial purity. The embracing of Aryanism, and the grandeur of Persia's past, was buttressed by archaeological ventures and the name change of the country from Persia to Iran, applauded by propaganda journals (*Iran'e Bastan*, Ancient Iran). There was a homogenization of the social façade by the enforcement of mandatory dress codes (Pahlavi hat for men, unveiling for women) and the emulation of Europeanism by the allowance of breathing space for women in education and social spheres in higher echelons of society.

All this was within the framework of a police state run by a military class (Cossacks, *Emir-Lashkars*) with vested interests in antiquated property relations in the countryside, serfdom and feudalism. This led to an assault on the nomadic tribes, not as eradication of patriarchal modes of exploitation, corvée (*bigari*), but as its monopolization by the ruling feudal-military-ecclesiastical and bureaucratic elite.

In the same vein, the political economy rested on landlordism, oil rent, and incipient industry, with oppressive taxation, not of wealth or income, but of consumption of the most basic staples (tea, sugar, tobacco, opium), to finance a burgeoning bureaucracy, within a caste society run as a caserne by an all-powerful Cossack with a secured bloodline.

The net result was that during the interwar years, the relatively stable period (mid-1920s to late-1930s), in all major categories of macroeconomic development (agriculture, industry, transportation, education), and without the benefit of Persia's petroleum riches, Turkey and Egypt outperformed Persia by a large margin. This is based on growth figures in these categories, spanning a period of fifteen years, and accounting for population differences, and this by the testimony of economic statistics compiled by an American economic historian, Professor Charles Issawi.[1]

Admittedly, the architecting of this system was as much the work of Teymurtash as Reza Shah.[2] It may even be argued that the Court Minister had far more to do with its conception than the king, who was principally in charge of execution (hence, the wobbly nature of Reza Shah's regime after the elimination of Court Minister in 1933 to its final collapse in 1941).

But for those shouldering the burden of this regime, the issue of authorship of the Pahlavi blueprint and how it had come to acquire this pompous name, whether it was the work of Teymurtash or the Shah himself, was immaterial. At the grassroots level, all suffered from this regime—subsisting as serfs (*ra'i'yat*), persecuted nomads (*ashayer*), downtrodden urbanites, and intimidated working classes. The working class suffered spectacularly: shattered trade unions and syndicates, the APOC strike broken at gunpoint, the latter so violent that it made a lifelong advocate of the labor cause out of Reginald Bridgeman, who abandoned his promising career in British Foreign Service and became a pro-labor political activist. The middle classes were squeezed as well by the burdens of taxation and bouts of inflation, and the intellectuals were demoralized and disaffected.[3]

The Court Minister, before going down on charges of rigging the exchange market with the help of German governor of the Central Bank, Herr Kurt Lindenblatt, had a vision of creating a national identity for Persia. The charges were valid according to the Court Minister, yet a pretext to eliminate him, as according to him the Shah was also a beneficiary in these schemes, at a much larger scale. The fall was preceded by the mysterious suicide in Beirut of Herr Volger, Lindenblatt's deputy, who had run away, entangled in speculation schemes in private banks in Europe, transacted on behalf of the Shah.[4]

At any rate, the Court Minister undertook a vast program of building mausoleums of the renowned savants and poets of Persia—Ferdowsi, Khayyam, Attar, Saadi, Hafez, and so on. He financed Persian orientalist work in Europe, the scholar Mohammad Qazvini a beneficiary. Likewise, at home, Mohammad-Taghi Bahar, the poet laureate, enjoyed the windfall, in both literary and political terms. He was parachuted into the *Majles* through the patronage of the Court Minister. The Court Minister also took a keen interest in archaeology, though this was another avenue for collecting illicit concession fees, playing the Germans against the French, the latter finally winning out (Godard over Herzfeld), but it is undeniable that he had a genuine interest in this project.

This was paradoxical, as he had attended boarding school in Russia since the age of eleven. When he returned to Persia, as a young man of twenty two, his Persian was rather rusty. He had to take private lessons to overcome his deficiencies, yet, he had now become the champion of the great literary tradition of Persia. Still more ironic, while he master-minded the millennium anniversary of Persia's national poet, Ferdowsi, by the time this took place, he was not only stripped of his power, but eliminated in *Qasr* prison by poisoning and suffocation. (*Qasr* bizarrely means "the palace," alluding to the origin of the site, mansion of a Qajar magnate.) This is somewhat reminiscent of the fate of Amir Kabir, chan-cellor under Nasser-ol-Din Shah, who had masterminded the creation of *Dar-ol-Fonun*, the Academy, but by the time of its inauguration, he had been murdered by court sanction (1852).

Now in 1934, a raft of European orientalists and cultural ambassadors came to Persia for the celebration of Ferdowsi's millennium, the Homer of Persia. This included the renowned Arthur Emanuel Christensen, the preeminent expert of Sassanid Persia, the great Dane who found in Reza Shah the incarnation of Persia's greatness and nobleness of its race.[5]

Taghi Erani could no longer bear the situation. As both Persian and a connoisseur of orientalism, with firsthand knowledge of how this racially tainted mind-set had degenerated into barbarism in Europe, he found the millennium celebration so repugnant, that he could not hold back his punches. The crux of his attack appeared in a single article in 1935, along the following narrative.[6]

If the Arabs adulterated the purity of the Persian race, did the Persians not commit the same toward those who thrived on this plateau before their arrival? Or are we to assume that this was a clean slate before the arrival of the Persians? If so, what to make of the Elamites, a dark-skinned race, in trade relationship with Dravidians, with a history stretch-ing over two millennia, before the earliest inklings of the Persians? What is the meaning of pure race anyway?

And if the Arabs were a part of the Persian Empire—we all know the story of Emir of Yemen, paying homage to Emperor Khosrow Anushirvan, and Ctesiphon, the capital of Sassanid Empire, touching Arab territories—then why should they be considered as aliens? Can one claim one, being an empire, and disclaim the other, its subject peoples, without suffering contradiction?

Further, if the Persian imperial regime was so magnificent, why did it collapse in the face of assault by hordes of barbarians? Can everything be attributed to the power of the sword? Had they not forged a creed at home (Arabian Peninsula) imbued with a sense of social equitableness (civil and property rights for slaves, tithe and almsgiving, public purse, *Beyt-ol-Maal*)? Had they not arrived not only with the sword but also with the banner of this creed born of social strife?

And is it not true that the empire had become devoid of its sense of social justice, operating as an oppressive band of aristocracy and priesthood, implanting fears of the afterlife in the minds of its subjects—serfs, slaves, vassals of varied ethnicities—while exacting tribute from them? How else to interpret the moral and political movements of Mauni and Mazdak, third and sixth centuries, before the arrival of the Arabs (seventh century), with deep yearnings for social justice? These were failed attempts at Zoroastrian reformation, succumbing to persecution and carnage, instigated by Magi priesthood.

And after the collapse of the imperial order and its dilapidated ideology, Zoroastrianism sapped of its original vitality (*Gathas*, early *Avesta*), by oppressive scriptures (*Vendidad*, late *Avesta*), was not the Persian renaissance (tenth century) born from power-grab by regional warlords plotting at the periphery of the Abbasid Empire?

Did not the national poet, Ferdowsi, offer his monumental literary *oeuvre*, epic of the kings (*Shahnameh*), to garner gold from the Turkish warlord (Sultan Mahmoud of Ghazni) with a foothold on the periphery, and designs on the entirety of Persia? Did not the same national poet vacillate to seduce Arab overlords by versification of the story of Yusef and Zuleikha (Joseph and Potiphar's wife), when he received only pittance from the Turkish sultan? The latter is so embarrassing that Persia's chauvinists deny that this was the work of the great poet.

Is not the same subservience to overlords unfolding before our eyes today, bowing to those in quest of worldly gain and dominance? Is it not this the real menace, and not the fiction of an adulterated race? Would not the achievements of modern science suffice to ensure progress, were it not for such retrograde views?

We conclude this section on the political mind-set of Taghi Erani by excerpts of his writings that shed light on some of the themes we have waded through:

> Our chauvinists mimic the ways of their fascist brethren, and wish to obliterate a whole swathe of history. Fascists claim they can wipe off the post-war years from their history, and pick up where they left off at the end

of the previous war....Likewise our chauvinists say we will purge anything that is not pure from our literature; we will even dispense with Saadi, as Arabic has crept into his prose and poetry....Since there the fascists are against Jews, here we must be against Arabs, and refrain from using their diction....Can there be any uttering more vacuous than that?[7]

Racial theory claims that disparity of civilizations stems from racial differences...black and red races are at the bottom, yellow higher, white highest. Amongst the white race, Aryans are higher than Semites. Amongst Aryans, Romans higher than Slavs, Germans higher than Romans. Although if the racial theorist is Italian, then the Romans occupy the highest rank. And within the noblest race there is yet more hierarchy; laboring classes the lowest, elites the highest...history refutes these fantasies... when Kushite and Mayans were thriving, the most civilized races of today lived in barbarity; civilization of so-called yellow race in China predates that of the white race. Amongst the white race, civilization first flowered amongst the Semites; only an imbecile can accept that a Jew has less aptitude than a German; that Turks are inferior to Bulgarians, one Asiatic the other European, and by the same logic, Japanese civilization is inferior to the Portuguese.[8]

Modern civilization has reached a dead-end; unemployment and social devastation are destroying the drive to invent and create; economic activity in recent years has been exclusively for purposes of war and destruction; a new foundation must be laid...In modern society, the majority seem to bear only responsibilities, while a minority is entitled to all the rights; rights and responsibilities must be bestowed on all members of society.[9]

POLITICAL PLATFORM

From Taghi Erani's political philosophy flowed his political platform and plan of action, and envisaged modes and methods of execution. In politics, as in science and medicine, so it seemed to him, the formulation of the problem is half of the solution, as it is the diagnosis of the problem that contains the germs of the solution.

To get a sense of Taghi Erani's views in this respect, what was for him the reality of political life in Persia and the political program that fell out of this, we have no better place to turn than the hundred or so pages of the political pamphlets of the CRRP. This is the covert group he co-founded in Berlin, with four collaborators, two of whom returned to apolitical life sometime after his return to Persia.

In terms of the specific events in Persia, notably grassroots political movements after Erani's return to Persia, it is not easy to get a good sense from the official press given the draconian system of censorship put in place and the intimidation of press chiefs. There is, however, a source which is germane to our discussion. This was an expatriate journal called *Peykar* (Combat) run by what seemed to be a coalition of activists and intellectuals from the left, including one of the co-founders of CRRP (M. Alavi). A total of fifteen issues came out, published in Berlin, then Vienna, during 1931 and 1932.[10]

It is improbable that Taghi Erani had any contribution to this journal. Some of its contributors were highly critical of the CRRP platform, finding it too mild. However, there was a strand more favorable to CRRP, at least as a potential partner in a broad coalition of the left. Their cacophony, as well as journalistic style—more blows than hard analysis—makes it unlikely that Taghi Erani was associated with it. Nonetheless, he did maintain contact with his former colleague in Berlin, who was collaborating with this journal. And its content echoed news appearing in the editorial and social columns of provincial newspapers and gave eyewitness accounts circulating in the country. This source provides important clues about the political climate in the early 1930s Persia from the perspective of the left, to which Erani belonged.

After the collapse of this journal came the appearance of Taghi Erani's own journal in Persia (*Donya*, twelve issues published from January 1934 to September 1935, four hundred pages, with several collaborators). Though the thrust of this journal was scientific and philosophical, its essays did have political undertones. Several issues had a lead column commenting on the domestic sociocultural scene. All issues had an editorial column on world affairs, which was explicit about economic and political developments in Europe: rearmament, fascism, appeasement, the United Front, and the rest. It also had a book column with snippets about the latest releases in several European languages, which echoed some of these themes.

Finally, there is the pamphlet of May 1, 1936 that was attributed to Erani by the police and was one of the charges brought against him. We also have access to his later views, expounded at the trial of the "fifty-three" (November 1938), a pamphlet of thirty pages, a truncated account of his defense plea which lasted several hours. This gives an important anchor to have a better read of his political mind-set in late 1930s. There are also remarks attributed to him by prisonmates who published their memoires. These evoke conversations that took place both before and after his arrest.[11]

***

The picture that emerges from these sources in terms of Erani's political orientation is as follows: Persia is medieval in substance, modern in external appearance; the court's annual expenditure one *Kror* (half a million *Tuman*, equivalent to half a million dollars, or a hundred thousand sterling), while the country's total budget is only fifty *Kror*, of which twenty is spent on the army, this for a population of twelve million. It is said that the court of Louis XVI was similar; it consumed about two percent of the national budget. The budget for health and education put together, less than the court's budget, while the country is in the grip of epidemics and illiteracy.[12]

In principle, the people have the right to representation. There is a *Majles* with deputies elected every so many years (typically two). In reality, these seats are up for sale. A certain deputy from Mash'had, who ended up as Persia's emissary to Berlin (Younes Vahafov), is rumored to have paid several hundred *Tumans* to Teymurtash to be put on court-approved list of *Majles* deputies. Before each election, the Court Minister dispatches such a list to the provincial governors, and it is understood that they must ensure these candidates come through. The candidates have annual salaries of several hundred *Tuman*, far more than a French, German, English, or American member of parliament, not even accounting for differences in standard of living, just as the king's official intake (a fraction of his total intake) is far more than the remuneration of Presidents or Chancellors of these countries with economic output at least one thousand times that of Persia.[13]

The policy of extracting ransom from tribal chiefs has too many tales to recount. In one incident in Azerbaijan, a certain Kurdish tribe who had proven unwieldy found their wives kidnapped by the Cossacks, with the message, come give your arm and get your wife back. This did not go down well, bloodshed ensued, claiming the lives of fifty Cossacks and untold number of Kurds. When Mirza Yahya went to collect the landlord share at harvest (typically four-fifths of total produce) from his villages near Mash'had, he was murdered. In the vicinity of Shiraz, a certain Red Mehdi and in Rasht a certain Heydar Khan have turned rogue, murdering the *mobasher* (landlord deputy) and fleeing to wilderness. Similar news came from villages of Hezar'e, Firoozak, Shabestar, Savad-Kuh, latter the birthplace of Reza Shah. Arson, murder, hostage, suicide are recurring themes.[14]

In Isfahan, Mirza Mohammad-Jafar runs a factory (*Vatan*, Nation) of five hundred workers—men, women, and children—with the most modern textile machines (Junker, thirty-six units, German). Women get one *Kran* a day (one-tenth of a *Tuman*), children a quarter of that (fifteen *Shahi*), and the most able male worker several *Krans*. Penalties for tardiness, poor workmanship, and indiscipline wipe off a good chunk of these wages, while the workday is twelve to fourteen hours, with no provision for scheduled rest, except a short lunch; nor is there a proper eating place, work uniform, sick leave, disability or old-age pension. In winters, children sleep there in makeshift cardboards, for fear of returning home at night. The wage-giver (*hoghoogh-pardaz*) and treasurer (*sandoogh-dar*), the latter a *Dar-ol-Fonun* alumni, have vindictive behavior toward the workers. In the south of the country, some twenty thousand suffer similar fate at APOC oilfields and refinery, segregation engineered by the company between Persian workers and immigrant workers brought in from India. Discipline is enforced by *sekurite-ofis*, gunboats showing up in response to strike action. A leading labor activist is murdered by the police (Morteza Hejazi).[15]

Yellow journalism abounds; stories of crime and suicide are a favorite topic of societal columns. They attribute this to modernization, collateral damage resulting from dark European novels, now plentiful due to the growing translation market. There is no mention of social malaise as the root cause of these tragedies. There is modernity for a minority. There is a sharp rise in the number of automobiles and sprawling mansions in the northern suburbs and social clubs. The moat is eradicated, but not the moat-dwellers. Citadel and historic gates have come tumbling down.[16]

The government is upbeat that there is robust revenue inflow. This comes from taxation, all of which is on consumption, and from tariffs (*Gomrok* or customs, on both import and export items), and royalties from oil. The latter are diverted to foreign accounts, a part of which is used for purchases of military equipment. The home budget is thus balanced without recourse to oil revenue. This seems brilliant except that the reserves are siphoned off to coffers of the royal family. With the collapse of the world economy after the 1929 crash, and nose-diving of Persia's exports, also collapsed the revenue from tariffs. This intensified the burden of taxation, which is levied not on wealth or income but on consumption, a basket including tea, sugar, matches, tobacco, and opium. Finance Minister (Taghizadeh) has boasted that there has been doubling of revenue from the latter source; more opium use, a real sign of progress.[17]

Central bank mops up the country's savings, but the balance sheet of the bank indicates over half of its reserves is invested in gilt-edged securities. For a country in dire need of infrastructure investment, savings are handed over to London for a guaranteed rate of interest, exceeding that accrued to depositors, difference siphoned off by bank barons and their political patrons.[18]

Government extended an invitation to Rabindranath Tagore to visit Persia, the spiritual leader of India's independence movement. His visit to be exploited as proof of Persia's strides toward sovereignty, a plea against this voyage by a renowned nationalist poet in exile (Lahuti), possible cause of it not materializing. Others, however, have fallen for such ploys, notably Professor Christensen, and the pompous millennium anniversary of Ferdowsi. Despite the reality of things, a positive image is fabricated.[19]

***

The solution sketched by the CRRP—recalling that this was a group of five, one a student of economics (Morteza Alavi), the others, students of political science (Ahmad Asadi), fundamental science (Taghi Erani), medicine (Ahmad Farhad), and law (Mahmoud Pour-Reza)—is a dual-pronged approach to eradicate landlordism and colonialism in one blow, spurring domestic industry with heavy-handed state intervention. Whether they knew this or not, this is a reiteration of a formula invented by a radical revolutionary of the constitutional era (Heydar Amoughli), when he expounded that in Persia the national question (colonialism) and the social question (feudalism) are intertwined and must be knocked out simultaneously; "national and social are two sides of the same coin," is the formula he minted.[20]

What the CRRP added to this formula is the notion of an industrial society in which the state plays a prominent role, and the notion of a political party as the catalyst to bring about this transformation. The former is largely German in conception (managed capitalism, with elements of coordination and protectionism, Rathenau its latest advocate in his *New Economy*), while the latter is decidedly Russian (theory of vanguard party, Lenin its architect). Oddly, Taghi Erani had found Lenin's analysis of modern science and philosophy (articulated in Lenin's *Materialism and Empirio-Criticism*) rather inadequate, but he does not seem to have followed through on this intuition in the realm of political practice.[21] Although Lenin is nowhere named in the political program, they suffered from a blind spot with respect to things Soviet.

Although Taghi Erani did shed this illusion in the latter stage of his life, while in captivity, as evidenced in his interrogation record, defense plea and remarks attributed to him in various memoires, by then it was too late, and this lapse in critical thinking arguably cost him his life. The CRRP program also falls short on implementation details.

According to the formula of inseparability of national and social questions for colonial and semicolonial countries, Persia falling within the latter category (in contrast to say India), the fundamental concept is that the enemy is within one's abode, and it is because of this internal enemy that the external predators are able to pounce on the country and keep it in a state of chronic underdevelopment and dependency and dark-ageism.

It is interesting how Gandhi began with the same logic but drew a radically different conclusion in terms of method. Since it is the cooperation of the Indian elites that enables British rule, it is by shattering this cooperation that India can dispense with British rule. Hence, his theory of noncooperation through the mobilization of the masses by levers of symbolism at the highest level and conflict at the lowest level, emphasizing nonviolence. In the context of Persia, this would be analogous to a *fatwa* against cooperation (what had led to the repeal of the tobacco concession in 1892, the dress rehearsal for the constitutional revolution of 1906).

In the case of Persia, this internal enemy manifested itself in a variety of forms—it appeared in the garb of orthodox religious authority, notably vast landownerships of religious foundations; endowments (*o'ghaf*) managed by an ecclesiastical order with which the Qajar and Pahlavi made pacts of non-aggression. It also appeared in the garb of the Khans and tribal chiefs ruling over as many as a million subjects engaged in agriculture, domestic industry, and pastoralism, laying claim to vast swathes of territory as ancestral lands. This was a serious bone of contention between the most powerful clans and the court. Likewise, it appeared in the garb of the king and the court, appropriating the most lucrative estates, the crown lands (*khales'e*). Finally, it appeared in the garb of absentee landlords, merchants and military chiefs and barons of bureaucracy managing their estates through intermediaries, a hundred hamlets (*sad-abadi, sad-paarcheh*) a common yardstick of ownership for this class of *nouveau riche*.

A radical land reform is the most essential ingredient of a progressive political program. This includes the distributing of the land to those who cultivate it, family holdings or cooperatives, and assisting them with machinery and loans through an agricultural bank, to eradicate serfdom and medieval exploitation. This boosts agricultural productivity and liberates productive forces for the development of industry. Reform also requires raising import tariffs to create a market for domestic industry and price controls to curb cost of living and stimulate savings and infrastructure investment via the deployment of savings; and it requires progressive taxation on wealth and income, solicitation of foreign direct investment and trade deals. Economic policy can be further buttressed by labor laws recognizing the right to unionize and banning current abuses, notably the exploitation of children and women as exemplified in the carpet industry.[22]

The goal is to develop a political and social policy consisting of separation of state and religion. As well, the separation of powers is imperative. Also, the executive branch must be stripped of extravagant privileges of the monarchic era. Ethnic minorities and nationalities are to be managed through a federal framework, a republic with a citizen's army for self-defense to ensure national sovereignty and a pacifist philosophy, critical in the current global climate, marching toward war. The introduction of universal education, vaccination and health services and social insurance are other important pillars. The banning of the mandatory dress code (veiling and unveiling) ensuring individual rights was key. In addition, constitutional guarantees for freedom of association, press, religion, political parties, elections, universal suffrage and abolition of discrimination (gender, religion, ethnicity), and harmful activities (opium production) all needed to be addressed.[23]

From this program, two questions arise—how did the CRRP envisage to bring about its ambitious vision, by what means and methods? And why did this program not win unanimity amongst forces of the left? The short answer to the latter is that because a majority considered that the nature of the program precluded its realization; that it was beset by an internal contradiction which made it utopian. And this was not because it was too radical, but because it was not radical enough. The program counted on domestic industrialists and merchants and financiers to play ball and not sabotage the system, whereas recent history had demonstrated the opposite behavior from these social classes, in Persia and even in Europe.[24]

What discredited this critique, as cogent as it seemed, for it captured an important element of risk concerning the fickleness of the elite classes in developing countries, was that it came from quarters that sought alignment with the Soviet Union and its prevailing foreign policy. And as this policy had a number of swings, the critique vacillated between charges of insufficiency of radicalism to its excess, back to insufficiency, back to excess, as mirrored in Comintern's seesaw, and so it lost credibility.

At any rate, Taghi Erani does not seem to have concurred with this critique, and as far as we can infer from the bibliographic record, he stuck with the original CRRP political line, and in the CRRP literature, we find only two clues concerning the question of means and method (we find many more when we examine his political practice)—one, the recognition that root cause of failure of Persia's nationalist movements in the last three decades has been the absence of a revolutionary party to rally the progressive forces of society: the working classes, the middle classes, and the intelligentsia, notably the youth. And two, the example of China where revolutionary forces are combatting not only feudal war-lords but also colonial powers, targeting the social-national knot alluded to above, making inroads toward establishment of people's power in the form of a revolutionary republic. This example from Asia was so stark, that the CRRP founders named their organization after its stated politi-cal platform (Revolutionary Republic). At least that is our best guess as to why they opted for that particular name.[25]

Thus, it was in this space that percolated Taghi Erani's thoughts, which sheds light on his mode of conduct during the eight-year period, from his return home to his arrest.

MODE OF EXECUTION, FORMULA FOR SEIZURE OF POWER

Working from the inside to effectuate change had always been a part of the political culture of Persia: Teymurtash under Pahlavi; Ghaem-Magham and Amir Kabir under Qajar; Nezam-ol-Molk and Rashid-al-Din (Hamedani) under Turco-Mongol rulers; Mauni and Mazdak under Sassanid emperors; and Zarathustra under a ruler of pre-Achaemenid times. All had attempted to infuse political power with the finesse of their intellects, capturing in part or in whole the essence of what the times mandated as necessity, for a range of motives, altruistic to egotistical, with mixed results.

Taghi Erani does not seem to have excluded elements of this approach, as his appointment to post of Vice-Minister of Industry, a ten-ure that came to a premature end with his downfall, lasting only a year (June 1936–May 1937), was not simply an act of dissimulation, as he did real work: a two-year plan drawn up to replace the German experts with natives in his administration. As well, he engaged in the evaluation of mining prospects as reflected in the photographic record, and perse-vered in the dissemination of scientific knowledge. He pursued all this with zeal and fervor.[26]

He also joined the local chapter of advocates of the League of Nations, covering his tracks may have been a motive, although he had never excluded legal activity from his political program, parliamentarianism included. The fundamental goal, however, did not change—to bring about a radically different social and political order, and this could not be achieved by other than radical means, though legal means could be used in a tactical sense: worker's cooperatives (welfare fund, print-setters), professional associations (doctors for social well-being), and social clubs (promoting sports, healthy lifestyle, particularly for youth).[27]

But there was a limit of what could be achieved by overt action, possibly a means to identify those with potential for more serious political engagement. The fate of the notables who had committed to working exclusively within the system was the best proof of the limitation of this approach. The sages of the bygone era may have sensed in the sovereign, the potential to act benevolently, an inclination the ruler may have had for some time due to political vulnerability. But invariably, the sages fell victim to their master or to antagonists of political power, perceiving them as traitors and collaborators.

The maneuvering space between these extremes, vacillating patrons prone to tyranny and determined opposition prone to vengeance, was virtually nonexistent. The insider paradigm had its limits. Covert action and confrontation with political power therefore were indispensable to bring about the required change; radical means were unavoidable.

This could be done in two ways, one resting on religion, the other on secularism, as its ideological pillar; Persia's recent history has provided ample examples of both. The net result was that secularism seemed to be the way to go, for it was in sync with the spirit of the time, but in a different way than it had been attempted up to that time.

The non-secular route seemed to have reached its limits. It was spent and exhausted. Even Khiabani did not attempt it as the rallying cry of his insurgency in Azerbaijan (1920), despite the fact that he was an accomplished clergy and superb orator, habile in addressing the common folks, by and large illiterate. He stressed the national spirit, as the thrust of his movement was against the Anglo-Persian Agreement (1919), Lord Curzon's attempt to turn Persia into a British protectorate, aided by his Persian accomplice, the Prime Minister, Vosough-ol-Doleh.

Kuchak, on the other hand, stressed Islam, but he had no inclination to revolutionary action, to dismantle landlordism. In his insurgency in Gilan (1915–1921), the breadbasket of Persia, rampant with destitute peasants, his action was pointed toward foreign intrusion (Russians and later the British). There seemed to be no voice within Islam, at that time, with a political punch, pointed towards the domestic order.

Islam was monopolized by an ecclesiastical order in alliance with political power—with the Safavid in the bygone era who made *Shi'a* Islam the official religion of Persia, a critical element of national coalescence, to withstand Ottoman expansionism, a *Sunni* power. The Safavid Shahs, notably Shah Abbas, tamed the *Shi'a* clergy with shrewdness and demonstration of extreme piety, journeying over a thousand kilometers on foot on domestic pilgrimage, Isfahan to Mash'had. Even in matters of religion and piety, the Shah stood higher than the highest ranking clergy. It was impossible to outbid him, making him the ultimate arbiter of all strategic matters concerning the state. And in those, he made moves with foresight, leading to a thriving economy, maritime and overland trade, flourishing culture, Isfahan its living witness.

And more recently, alliance with the Qajar—clerical order rocked the monarch only to keep him in check, but not topple him. The tobacco rebellion took place, for the king (Nasser-ol-Din Shah) had auctioned off too much to foreigners, intruding on ecclesiastical land tenure and privileges. And in the constitutional movement, his successor (Mozaffar-ol-din Shah) had neglected the consultative duty, both prime examples of conditional collusion, as was the earlier collusion to defeat the only serious threat to political and religious power in the garb of Islam, Babism (c. 1840s), advocating total dispensation with priesthood, Lutheran in spirit, which was so vehemently crushed that its sequel became apolitical, *Bahaism.*

And most recently, the alliance with Pahlavi—a deal in which Reza Khan would drop the republic idea and leave intact the system of ecclesiastical land tenure, the *o'ghaf*, and in exchange the clerical order, as epitomized by its leading authorities in the holy sites of Iraq, would not sabotage his ascension.[28]

When queen Jahan, Ahmad Shah's mother, journeyed to Iraq to urge the leading *Shi'a* clergy issue a *fatwa* against Reza Khan's power-grab, she was received with total indifference, as a new arrangement had been made, which sealed not only the fate of her son, the irresolute Ahmad Shah, but also that of Modarres, the flamboyant deputy, the last representative of the court-clergy pact under Qajar, buttressed by the backing of Bazaar, the commercial class, merchants and guilds.[29]

Therefore, those who could use Islam in a virulent way were an extreme minority (Bohlul, an Afghani cleric, with a following in

Khorasan), and in no position to mount a serious political challenge. Re-appropriation of Islam as a political weapon came much later, in 1960s, culminating in the revolution of 1979, and that because of repeated failures of secular attempts to tame the elitist state.

At the time, in 1920s and 1930s, secularism seemed to be the only sensible route, and as indicated, even Khiabani did not contemplate resorting to religion to articulate his radical political program. Radicalism in politics and Islam did not go together. Complicity and proximity with officialdom had discredited Islam as a viable political force. An alternative interpretation had not yet been invented. That would come much later with Ayatollah Khomeini.

Thus, what precluded Taghi Erani from evoking Islam for political aims (something which must have crossed his mind, as he even resorted to working from within the royalist system, so nothing was off limits) was not his lack of familiarity with this universe, for his maternal grandfather and great grandfather were notable theological scholars (*Mojtahed*). He, himself, had impressive knowledge of Islamic scriptures, attested to by a leading *Shi'a* theologian (Ayatollah Motahhari), and evoked by one of the principal power brokers of the Islamic Republic (Hashemi-Rafsanjani), but that at that particular juncture in Persia's history, radicalism and Islam did not go together.[30] Islam had become apolitical, pacified, compromised.

It is also quite possible that Taghi Erani's deeper reflections led him to believe that it was doubtful that the two would ever go together. Although there was in Islam a strong accent on social justice and equitableness, to which he alluded, it also maintained that private property is sacred. But lopsided amassment of private property, land its principal form, from which the vast majority were deprived, was the source of the evil. By binding the serfs to the land, not only were they exploited by their overlords, but it prevented them from turning into working-class material for the development of industry and capitalism, the foundation of modern society, manifesting as Europeanism in political, social and cultural spheres.

So, it was inconceivable how Islam, conservative by creed, certainly in economic terms, resisting redistribution of property, could hold the key to radical social change. What was in vogue at that time was Proudhon's formula that property is theft. Thus, unless this evil was put right, monopolization of land ownership and antiquated feudal relations, blatant slavery, safeguarded by Reza Khan and his military chiefs (*Emir-Lashkars*), nothing would advance.

It is therefore not surprising that the most serious challenges to the established order—Khiabani in Azerbaijan (1920), Ehsanollah Khan and Heydar Amoughli in Gilan (1920–1921), and Pasyan in Khorasan (1921)—were all brought down by the complicity of central power in Tehran, and provincial magnates and their mercenaries, and fellow travelers prone to capitulation. Foreign intervention played a role as well, but in the overall scheme of things it was not the principal factor. These movements were all secular, as for them there was nothing sacred about the maxim "property is sacred." Indeed, some made daring moves to redress the lopsided distribution of property and wealth.

The leaders of these insurrections had been crafty enough to seize political power in a major city or province, each with its own intriguing story, but they were not sturdy enough to hang on to that power and expand its base, as they were surrounded by hostile feudal warlords and Khans ready to pounce on them. The insurgents had found no way of toppling their adversaries by stirring riot and revolt amongst the exploited subjects of their rivals.

These movements each lasted six to twelve months, giving ample anecdote to why secularism is necessary, as it shatters the taboo attached to wealth and property, but it is not sufficient to guarantee success measured in terms of advance towards seizure of political power. Taghi Erani alludes to these failed secular nationalistic movements, and the singular idea that emerges is that unless there is an explicit class-conscious orientation and social content in the insurgency, it will succumb to hostile forces. To rally such movement, a class-conscious political party is needed, and to create such a party what is needed is a hardcore group of revolutionaries convinced that this mantra is the only winning formula, forming the backbone of the party, its cadres and leadership, preparing the ground until such time as social and psychological conditions ripen sufficiently to permit seizure of political power.[31]

Through such reasoning, Taghi Erani arrived at a formula for seizure of political power, on a par with what the Russians had minted. In their case, it had been a long evolution from the romanticism of Narodniks, going amongst the peasants in rustic attire to give a helping hand, to the terrorism of Social Revolutionaries (Lenin's brother emerged at the incipient phase of this movement with an attempted assassination of the Tsar to rock the masses), to the so-called professionalization of revolutionary action by the Bolsheviks. This was articulated by Lenin, a lawyer with a globalist vision, yet deeply nationalistic and pragmatic. Ultimately, he sacrificed everything for its sake, his avowed internationalism, the world revolution, and the right of nations for self-determination.[32]

And China seemed to be another manifestation of the same principle, in fact, one more akin to conditions of Persia, for feudalism loomed large and the country was colonized by competing powers, Japan in the north and Great Britain dominating in the south, analogous to the case of Persia with Russia and Great Britain hemming it from above and below.

The CRRP literature had explicit passages on China, its liberation movement, simultaneously taking on colonialism and feudalism (so referring to its communist faction). Evidently this had struck Taghi Erani and his cohorts as showing the way forward, but what they did not know is that what led to the success of the Chinese communist movement was in large part due to their rupture with Comintern. The devil was in the detail, and that detail they certainly did not know. Indeed, they tried to attract the attention of Comintern to their cause, soliciting collaboration via Louis Gibarti, Comintern liaison in Berlin. Gibarti's real name was Laszlo Dobos. He was Hungarian, and though attached to Comintern, worked most likely for Soviet intelligence. His mission was to keep an eye on Willi Münzenberg, his close collaborator, who was later taken out by a Comintern agent.[33]

The Comintern, utterly Russo-centric and pragmatic to the extreme, took the Chinese nationalist party into its membership, the Kuomintang or KMT, which was non-communist, and as it turned out, anti-communist. This is because the Russians saw them as having the best chances of defeating British and Japanese colonial rule in China, giving to Russia a useful ally.[34]

This policy was so systematically followed that even when the KMT launched an assault on its allies, the Chinese communists, in spring 1926 in Canton, they obliged the Chinese communists to continue their collaboration with the KMT. Thus, Comintern help came with strings attached. When the second assault happened, of much greater scale, spring 1927 in Shanghai, they still did not let go, and only switched from Chiang Kai-shek to another wing of the KMT as their ally. These events led Mao and his entourage to follow their own strategy and eventually win political power, largely independent of Russia and Comintern.[35]

A similar thing was done in the case of Great Britain. Comintern funded the Trades Union Congress (TUC), relegating the British Communist Party (BCP) to subordinate position, as they considered TUC

has the mass and muscle to bring into power a government with pro-Soviet foreign policy, whereas the BCP had no such might, being a fringe
party (in contrast to communist parties of Germany, France, Italy).[36]

Likewise, in the case of Persia, they piloted its communist forces as a
function of Russia's existential needs and relationship with Great Britain,
the hegemonic colonial power in the Near East. This went through two
cycles, each starting with confrontation and ending with *entente*: thus,
starting with directives to Persia's communists to stir agitation and confrontation with forces allied with Great Britain and ending with directives to hold back and lie low until further notice.

The first cycle spanned 1917–1921. It started when the Bolsheviks
needed to mobilize the Moslem ethnicities in its southern fringes to
help it ward off the assault by white Russian generals, backed by Great
Britain (which organized an invasion also from the north, Archangel, but
failed).[37]

The Persian peasantry that had emigrated to the Caucasus and
Turkmenistan, homesteaded there over decades, and had its own cultural and political organizations in Baku and Tashkent and other towns
was a prime target to be recruited into the cause. This led to the founding of *Edalat Party* (Justice Party, 1917, Baku), its rebranding as the
Communist Party of Persia a few years later and the infusion into the
Gilan insurgency, which had its own indigenous leftist elements in coalition with Islamic forces, a united front against foreign intrusion. The
high point was when the Bolsheviks arrived with their fleet at Enzeli,
loaded with arms and cannons, led by a couple of their political and military heavyweights (Ordzhonikidze, Stalin's associate, and Raskolnikov,
commander of Bolshevik fleet in the Baltic and Caspian), in May 1920.

This was a strong signal to the British to leave Persia (Norperforce,
several thousand strong), for their spillover to Soviet territory was considered imminent (as it had happened during the Great War with General
Dunsterville's forces arriving in Baku). Within a year or so this tug of
war led to an Anglo-Russian *entente*, and the installation of Reza Khan
in Tehran, engineered by General Ironside, and the departure of both
forces from Persia, consummated in official pacts.[38]

These events led to the eradication of Persia's communist leadership,
withdrawal of Soviet support leaving them exposed, a mix of execution
by local mercenaries and exile to Soviet Union. Those who remained
regrouped, but their role would be to keep a nominal organization,
partake in overt activity, social clubs and guilds, soft journalism and

parliamentarianism, and not undertake any subversive or radical action. Only those natives who were recruited in Russia's intelligence service would be pursuing a more focused line of action (recruit sympathizers in the armed forces or other government bodies and provincial players), for intelligence gathering, and future activation. This was the first cycle.

The second cycle spanned the period 1927–1935. It was triggered by the fallout between Great Britain and Russia, due to Russia's meddling in the British general strike of 1926: the formation of the League Against Imperialism, the anti-colonial conference of 1927, and support of anti-British forces in China and India precipitating it. These led to the rupture of diplomatic relations between the two countries. This happened in 1927. Although diplomatic relations were reinstated in 1929, reconciliation was slow, and it was not until 1935, at the seventh congress of the Comintern, assessing fascism to be the most serious existential threat to the Soviet Union, that is Hitler and Mussolini and Hirohito, that it adopted the policy of United Front.

This was a somersault from the sixth congress, held in 1928, at the height of Anglo-Russian hostility, which dictated that member parties adopt the most virulent strategies to challenge and overthrow their home governments, including the denunciation of any parties that contemplated parliamentarianism and legal work. Those were petit bourgeois parties, clinging on to their class illusions. Nothing less than total eradication of imperialism and its stooges would do; that was the rhetoric in 1928.

In 1935, the rhetoric had metamorphosed to its exact opposite, a United Front. This meant member parties in the zone of influence of so-called western liberal economies had to refrain from radical or revolutionary action, and if need be form coalitions with their home governments. This included Indians giving up the idea of shedding British colonial rule, Chinese trying to battle it out against British-supported forces of the KMT (for now the directive was to patch up old hostilities and fight off the Japanese), or the Egyptians and Syrians trying to do away with their indigenous dictatorships tied to British and French colonial rule.

In the case of Persia, the onset of the cycle, marked by confrontation, meant reactivating the communist network. This unleashed a series of events—convening the party's second congress (seven years after the first, 1920–1927), renouncing petit bourgeois parties (including the CRRP), and triggering industrial action to the extent possible (the APOC strike containing elements of this, and the textiles strike at Vatan). Whereas at the tail end of the cycle, marked by conciliation, the opposite directive was given. This transpired as—rebuilding the organization after police's crackdown (Agabekov defection triggering it), resorting to soft activism, liquidating any activity that may be misconstrued as challenging the regime in power, preaching the gospel of United Front, May Day anniversary an opportune moment as solidarity with the elite in power as a sign of national unity. In a nutshell, they were given a diktat of hibernation until the next activation. And there would be yet other activations and hibernations, but that falls outside the time scope of our narrative.

This second cycle coincides with the political career of Taghi Erani, from Berlin to Tehran, with excursions back to Berlin, Paris, London, and Moscow; by delving into his activities and following his footsteps, we develop a better understanding of the meaning and mechanics of this cycle, and by recognizing this cycle, we develop a better understanding of his mind-set and *modus operandi*, and how in the midst of this contradictory maelstrom he endeavored to leave his mark.

NOTES

1. See Issawi, *The Economic History of Iran*, pp. 375, 379, for the raw data based on UN and US statistics. Author has used these data to compute growth figures for the 1925–1938 period, accounting for population differences by normalization.

Categories of economic activity	1925–938 Growth in %, Iran vs. Turkey, Turkey = 100%, normalized to population	1925–1938 Growth in %, Iran vs. Egypt, Egypt = 100%, normalized to population
Railroads construction, km	61	153
Automobiles in circulation	305	73
Cement production, tons	31	25
Refined sugar production, tons	10	19
Cereal and rice production, tons	53	94
Cotton production, tons	72	9
Energy consumption, tons of coal	78	84
Students in primary and secondary	44	26

2. See Motevali, *Teymurtash va Baziye Ghodrat*, pp. 22–23, for Reza Shah's address to Teymurtash: "Me the army, you the state."

3. See Bridgeman, *Persia and British Labour, July 1931*, also see Cronin, *Popular Politics*, on the 1929 APOC Strike. For demoralization of intellectuals, see Aryanpour, *Az Saba ta Nima*, p. 313: on "band of suicide" (*band'e khod-koshi*) amongst dramatists (the playwright Reza Kamal alias Shahrzad, as well Mojtaba Tabataba'i, Seyyed Reza Khan Sadr, Habib Meykade, all committed suicide); also see Shafie-Kadkani, *Advar'e She'ere Farsi*, p. 54, on despondency (*enzevagari va sarkhordegi*) in the literature during the Reza Shah era.

4. See Majd, *Great Britain and Reza Shah*, p. 180, for Teymurtash's assertion that he stole but the Shah stole even more, as he was the principal beneficiary; see Najmi Alavi, *Sargozasht'e Morteza Alavi*, pp. 146–148, for fleeing of Volger reported in *Peykar*; see Motevali, *Teymurtash va Baziye Ghodrat*, pp. 74–76, 101, 103–104, on the fate of Volger and Lindenblatt. Also see Khatib-Shahidi, *German Foreign Policy Toward Iran Before World War Two*.

5. See Asmussen on *Arthur Emanuel Christensen* and his praise for Reza Shah. He made three trips to Persia (1914, 1929, 1934), last one for the millennium celebration of Ferdowsi.

6. See *Taghir'e Zaban'e Farsi* also *Materializm'e Dialektik* (in Maghalat, pp. 37–38), and *Bashar Az Nazar'e Maadi* (in *Donya*, 12-issue Volume, p. 329, article spanning pp. 300–366).

7. See Erani, *Taghir'e Zaban'e Farsi*.

8. See Erani, 12-issue Volume, p. 329.

9. Ibid., pp. 309, 357.

10. See partial reprints of *Peykar* in Najmi Alavi, *Sargozasht'e Morteza Alavi*; intimidation of press chiefs described in pp. 50–52.

11. Prisonmates who published their memories and make references to Taghi Erani include: Alavi, Alamuti, Avanesian, Boghrati, Eftekhari, Eskandari, Gorgani, Jahan-Shahlu-Afshar, Maleki, Pishevari, and Tabari.

12. Summary extracted from the CRRP manifesto, *Bayan'e Hagh* (1927), which examines the Reza Shah regime through the lens of national budget and political economy.

13. See *Peykar*, reproduced in Najmi Alavi, *Sargozasht'e Morteza Alavi*, for details of corruption allegations.

14. Ibid., pp. 54–55, 70, 126; peasant uprisings.

15. Ibid., pp. 59–60, 63–65 (Reginald Bridgeman article), 103–105, 117–120; worker's conditions.

16. Ibid., pp. 61–62; suicides.

17. Ibid., pp. 135, 146–148; financial irregularities. For Taghizadeh's boast on opium revenue growth, see Afshar and Dehbashi, *Zendegi'e Toufani*, p. 223.

18. See Najmi Alavi, *Sargozasht'e Morteza Alavi*, pp. 139–140; Bank Melli.
19. Ibid., pp. 115–116; Tagore.
20. See Rais-Niya, *Heydar Amoughli*, pp. 221–222: "national revolution must be transformed to social revolution"; elaborated on pp. 313–317.
21. See Erani, *Materializm'e Dialektik* (in Maghalat, p. 26), for assertion that Lenin's *Materialism and Empirio-Criticism* is not up-to-date on modern science.
22. See *CRRP Political Program*, March 29, 1926.
23. Ibid.
24. See critique of CRRP political program by Sultanzadeh in Ahmadi, *Tarikhche Ferq'e Jomhooriye Enghelabi'e Iran*, pp. 22–23.
25. See CRRP, *Statement of Anti-Colonial Stance*, January 19, 1928, on the principal cause of failure of political movements in Iran, lack of a vanguard political party.
26. See Momeni, *Parvandeh*, p. 285, for Erani's description of how he developed a two-year plan for local experts to take over from German experts; how he ran his office with discipline, for which he became reputed; his mine visits are reflected in the photographic record (*Erani Family Archives*), and scattered comments in his writings (e.g., *Donya*, 12-issue Volume, p. 375, article on *Taghir'e Zaban'e Farsi*).
27. See Momeni, *Parvandeh*, pp. 91–92, 239 for summary of Erani's overt actions. Also see Ahmadi, *Tarikhche Ferq'e Jomhooriye Enghelabi'e Iran*, pp. 61–62.
28. See Cronin (ed.), *The Making of Modern Iran*, p. 21, in article by H. Katouzian and how Reza Shah dropped the republic idea after meeting with clerical authorities in Qom (Hajj Mirza Husayn Na'ini and Sayyid Abulhasan Isfahani). His assault on the Imam Reza shrine (July 1935) was a violation of this pact (fall of Mohammad-Vali Asadi).
29. For an account of queen Jahan's journey to Iraq, based on memoires of her grand-daughter, see Pasyan and Motazed, *Zendegi'e Reza Shah*, pp. 239–240.
30. In the preface to the work of Aytatollah Tabataba'i, *Osool'e Falsaf'e va Ravesh'e Realizm*, written by Aytollah Motahhari, there are numerous references to Taghi Erani and that his propounding of the philosophy of materialism in the Iranian cultural context is best due to his familiarity with Islamic scriptures and Arabic. Also see allusion to aura of Taghi Erani in the 1940s by Hashemi-Rafsanjani, in *Rishe'ye Ghodrat'e Enghelab*; both references came to my attention through the work of Ahmadi, *Tarikhche Ferq'e Jomhooriye Enghelabi'e Iran*.
31. See *CRRP Political Program*, March 29, 1926.
32. Lenin's essay on the *Right of Nations to Self-Determination* was published in 1914; but in the Russian civil war (1917–1922), when Lenin was fully

in charge, the Bolsheviks denied independence to the Caucasian republics (Azerbaijan, Armenia, Georgia), and Turkmenistan.

33. See *CRRP Statement of Anti-Colonial Stance*, January 19, 1928, for praise of the Chinese revolutionary movement.
34. See Hallas, *The Comintern*, pp. 120–125.
35. Ibid.
36. Ibid., pp. 118–120.
37. See Tohidi, *Matbooat'e Komonisti'e Iran dar Mohajerat*, also Pishevari, *Tarikhche Hezb'e Edalat*, and Rais-Niya, *Heydar Amoughli*, for how the Bolsheviks recruited Persian immigrants in Caucasus and Turkmenistan, via Edalat Party, mobilizing them in the Russian civil war, against whites. Asadollah Ghafar-Zadeh, leader of Edalat, went to Gilan to expand this movement into Persia (he was identified by the police and murdered, 1918). Rothstein and superiors up to Lenin (Karakhin, Litvinov, Chicherin) forged the policy of *entente* with Great Britain (1921); see Rothstein's offer to British legation in Mash'had to help crush the Pasyan uprising (October 1921), in Cronin, *An Experiment in Revolutionary Nationalism*, p. 731.
38. The Anglo-Soviet commercial treaty of March 1921 was in part the enshrining of the Anglo-Soviet *entente* over Persia; this was preceded by the Soviet-Persian treaty of friendship of February 1921 (which had been in the works since December 1920, and contained the clause which became pretext for Soviet invasion of Iran in August 1941).

CHAPTER 12

The Quiet Period (1929–1933)

Taghi Erani arrived in Tehran in January 1929. He had been on a government stipend during the last two years in Berlin. He had received sixty *Tumans* per month (the equivalent of sixty dollars, three hundred marks, twelve sterling) and his return journey was also paid for by the government.

He returned home either by rail to Russia, then Soviet Azerbaijan (Baku), a ferry to Enzeli, then Rasht, Qazvin, Tehran; or sailing from Marseilles to Constantinople and Batumi, then Tabriz, Qazvin, and Tehran; or sailing via the Suez Canal, arriving at Bushehr, then Shiraz, Isfahan, and Tehran. All three were well-oiled routes in those days. The fastest any of these could be done was about ten days, plenty of time to ponder over what had come to pass and what awaited him.

His financial support during those two years had come from the Ministry of War. Each ministry had a budget and supported its own students. They had specific fields of study of interest to them and preferred spots to dispatch their students. Taghi Erani had joined this scheme after the fact, as he had gone to Germany on his own account. The financial support meant that he had an obligation to serve, two years for every year of assistance, unless he wanted to pay back what he owed.

According to his schoolmate from the *Dar-ol-Fonun* days, Dr. Mehdi Azar (the Minister of Education under Mossadegh), who had brief encounters with Taghi Erani in the mid-1930s, Erani opted not to join the government service as a full-fledged employee and decided to pay

© The Author(s) 2019
Y. Jalali, *Taghi Erani, a Polymath in Interwar Berlin*,
https://doi.org/10.1007/978-3-319-97837-6_12

back his debt in installments. According to government records, drawn up when the "fifty-three" were arrested, and this by instructions from the king's office to collect all such fees, six having studied abroad, Taghi Erani had not paid back his debt. Either way, it meant that he did not join the government service upon his return, at an official or permanent capacity, as otherwise the debt would have drawn down.[1]

Also, it is a certainty he had debriefing about his activities in Berlin, possibly at the Ministry of War. A similar thing happened to a student who had returned from France, was debriefed at the Ministry of Justice, as he had been a law student and had dabbled in political activity.[2] Taghi Erani, by all indications, was cleared of wrongdoing, though he knew that he could be under surveillance for some time.

He then embarked on a career in teaching, which was a natural choice for him. He freelanced as a teacher of physics, chemistry, biology (botany), mathematics, and even German in several schools. All schools were under the jurisdiction of the Ministry of Education. This meant that nominally he was an employee of this ministry, with an assigned grade and compensation level. These details appear in his interrogation records.[3]

Two of the schools in which he taught were special schools, associated with the Ministry of War. He taught chemistry there, which was a field of interest to this ministry. He discontinued one after a short period but seems to have carried on with the other, and this perhaps gave him respite from paying back his debt immediately, possibly a grace period from the ministry, as long as he taught in one of their establishments or did ad hoc work for them.

Below, a note the author received from one of his students at the school where he taught for a brief period (*Madres'e Nezam*, Military School):

What I offer is not a document or manuscript, but perhaps knowing it would not be devoid of interest. That serene-soul (*shad'ravan*) taught us chemistry in the third year of *Nezam* high school. Presumably it was year 1308 (1929). Although his teaching did not last more than several sessions, in that short time we were drawn to him, due to his method of teaching and instructive allusions. I had sensed him to be patriotic and someone with new ideas. For instance, in one of his digressions he said we ought to adopt expressions in Persian for certain scientific terms, including in chemistry, as the Egyptians have done. Instead of 'sulfate de soda', they say 'kebritat'e sodium'. We can also say 'googerdat'e sodium'. Unfortunately, after several sessions, we no longer saw him.[4]

For Taghi Erani, the beauty of working in education was that he could be in touch with the younger generation during the formative years of their lives. He was eager to see youth from all walks of life and get a sense of their mind-set, issues, and circumstances. It also allowed him to develop his network among teachers and school administrators. This was a well-knit community (*farhangian*), and their profession gave them certain openness to new ideas and sense of social responsibility. It would permit him to resume his publication work in the scientific domain, and it gave him a great deal of freedom. He did not have much by way of administrative duties. He was in charge of his classes; he would teach, and on occasion proctor and grade diploma examinations for his precinct. He could forgo a good deal of official chores and office politics.

The downside was that the money was not great, but he did not have a family at his charge and lived in his mother's home, and this seemed to suit him much better than working in government bureaucracy. He also did ad hoc consulting work in industrial chemistry. He was retained by the company producing hygienic products for the consumer market, to propose a formula and process to produce body soap (commercialized as *Darugar*, using oils of date seed and olive seed, with ads in his journal). He also did consulting work for the Ministry of War, and at least one of his trips to Berlin was in the context of evaluating recent developments in forensic chemistry for criminal investigation.

He resumed his publishing work with a great deal of energy. The *Series in Exact Sciences* (*Selseleye Oloom'e Daghigh'e*) was mostly done during this period. He had produced his first volume of physics in Berlin. He reprinted this in Tehran, put out volume two, which was more voluminous than the first; and put out a treatise on modern physics, which can be considered as volume three, expounding basic concepts and mathematical expressions of the theory of relativity and quantum mechanics. It was called *Theories of Science* (*Teorihaye Elm*). He put out a compact one-volume book on chemistry. These he used in his teaching, and to some students he would give out free copies.

A university was also being formed during this period. This is not what we may imagine as a university today. It was essentially the second floor of *Dar-ol-Fonun* with a parallel administration, teaching those who had graduated from high school, in limited fields of science and engineering. Taghi Erani taught at the university, and some of his high school students ended up going there. His books on physics and chemistry were also used as textbooks, for there were no books in Persian in these domains at that time, particularly covering modern and applied science. He had plans to publish a book on biology, as it was a logical component of his series, also useful for medical students, but it does not appear that this project materialized.

He also went back to his psychology. He had published a book on individual psychology in Berlin. Now he expanded that work, adding to it a section on sociohistorical and cultural psychology and published it in Tehran.

These activities gave him a good feel for the publishing trade in Persia: the approval process, the publishing outfits, the presses, the print-setters, the booksellers, the buyers, the whole chain. Besides the above works, he announced the forthcoming publication of several essays, but these do not seem to have reached fruition.

In terms of his social life, he resumed contact with former acquaintances, got to know new people, a few of them aspiring artists in the theater and authors of fiction, connected with some of the Berlin crowd, and several who had studied in France. He brought with him a social practice which apparently was German. He would fix specific hours on a specific day of the week, and those with whom he was acquainted could drop by his home during this time and have casual conversations. These were his so-called office hours. He did this every Sunday evening, starting at seven o'clock. And with a few, he would meet on a one-to-one basis for more private conversations.

He maintained contact with Germany via correspondence. He received journals from abroad, *Rundschau* and *Journal de Moscou,* and picked up what he could from Brukheim, a local bookseller. He got absorbed in self-study, for he had brought a number of books from Germany on philosophy, economics, sociology, history, biology, and mathematics (non-Euclidean geometry). A partial list of this is found in the police records, who returned to his home several times (with a police translator, familiar with European languages) to search his papers, letters, and library.[5] These were at his office, which was one of the rooms in his mother's home converted for this purpose.

He had a habit of walking to whichever school or appointment he had to go to. He did not have an automobile and would not take the *droshky* which was common in those days. He had a cane, wore thick glasses, was generally well dressed, in a suit, barrel chested with a smallish frame, and while walking with his usual brisk pace, would observe the people, streets, shops and ponder about whatever crept into his mind. He describes this mode of contemplation, engaged in movement, largely unaware, in his *Psychology.*[6]

He would also venture out of town; to Varamin, known for its orchards; to Tajrish, in the foothills; the latter by mule-ride with pleasantries scribbled about some of these outings in his notes, how his mule misbehaved and would not obey him; and further afield, a voyage to Mazandaran, known for its forests and seashore, the Caspian.[7]

All in all, it was a period of calm and tranquility, perhaps just what he needed after those hectic years in Berlin and its erratic ending. This cadence persisted for some time, five years in fact (1929–1933), and during these years there was much that was changing in the world, especially the world that he knew intimately.

Moving Landscape—Germany, Russia, England, Persia

Germany was perhaps the biggest casualty of the 1929 crash, Black October. It does not seem to have had the capacity to withstand such a shock, to go from half a million to six million unemployed in just a few years. An economy strapped with debt (war reparations) and hugely dependent on exports which melted away overnight: hence, the collapse of Weimar, the rise of Hitler, and the disfigurement of the political landscape; January 1933 the official switch, Hitler at the helm, though the Nazis had been active since the early 1920s.

The Reichstag fire (February 1933) became a pretext to ban the opposition. The communists and socialists were outlawed. Most of its leaders fled to Sarre (a German territory under French rule) or to France; later they would get trapped there as well, when the French started colluding with Hitler (Vichy regime). The left seemed to be paying for the fatal errors of its past.

When Hermann Müller became chancellor, he banned May Day demonstrations to appeal to his right-wing coalition partners. But May Day took place anyway (1929), and when it turned violent and the police opened fire, dozens lost their life. There were barricades in Berlin, markedly in Wedding and Neukölln, with a concentration of working-class neighborhoods, chronicled in a chilling novel, *Barricades of Berlin*. Erani had walked through some of these neighborhoods, and mocked Alfred Rosenberg, the racial theorist, for never having set foot there, for how could he claim the Germans to be the superior race when there was nothing but misfortune and misery to be found there. Years earlier, Ebert had taken the socialists to war (only Karl Liebknecht dared oppose him, in his famous eight-word speech—"Down with the War, Down with the Government!"—before being dragged away by the police), giving to Germany maimed veterans visible on the streets every day, and two million fatalities, mostly of working-class background; and Müller continued the tradition, inventing the muzzled society.[8]

For these reasons, many on the left of the political spectrum turned to the communists, but they failed their test as well. Karl Liebknecht and Rosa Luxemburg were the last communist leaders who were beyond reproach, at least in terms of personal integrity. But they became a casualty of a failed revolution, which had surprised even them, Red Germany, the short-lived Berlin and Munich soviets, 1918–1919 (democracy through direct participation, soviets not parliament). This sprang spontaneously from the grassroots level, an expression of desperation with a ruling elite fronting for financiers and industrialists that had dragged them into the war for acquisition of territory and markets and could not even ensure a regular food supply to the population (chronic starvation, the turnip winter). After the failure of this revolution, crushed by right-wing militias (*Freikorps*), the German Communist Party (KPD) lost its bearings.

When the KPD signed up to the Comintern bylaws (twenty-one conditions, particularly number sixteen, the authority clause, mandatory for any party wishing to be a member, formulated in summer 1920 by Lenin), it meant that it was no longer an autonomous party, but subject to organizational dictates, money and resources from Moscow, but with strings attached, if not chains.[9]

The independent leadership of the KPD was sacked, and Thälmann took the helm. Then, they decided to go for a German October (1923, the peak of hyperinflation), Zinoviev running the show from Moscow, fixing the date of the insurrection, Kiel and Hamburg its focal points, dispatching arms and advisors and money. This backfired because it was engineered from abroad and the dockers and workers did not go along with it.[10] From there, the KPD went down the tube, leading to its total Stalinization.

Meanwhile, the Russians swiftly moved to "plan B." Denial that this failed *coup* was the work of the Soviet State (Comintern's international garb a convenient cover) and carried on with the Rapallo Treaty that they had signed a year earlier: secret deals with Weimar to build chemical weapons on Russian soil with German knowhow and financing, a win-win deal. The Russians would get the goods, the Germans a way to circumvent the straitjacket put on them by the Versailles treaty. Rathenau was the architect of the deal.[11]

At any rate, the German left, whether socialist (SPD) or communist (KPD) or in-between (USPD, Independent Social Democrats), was shrinking in electoral terms, at best trading votes between them, as the Nazis went from strength to strength, the 1929 crash giving them a big shot in the arm, and catapulting them to power in 1933. Now the left was outlawed, its leaders executed, locked up or on the run.

The political celebrities Taghi Erani knew, Willi Münzenberg (of the KPD, Reichstag deputy and organizer of the Brussels anti-colonial conference attended by Erani), and Rudolf Breitscheid (the USPD deputy who wrote to *Vorwärts* to prevent the expulsion of Erani's CRRP colleague), both became fugitives, ending up in France, and both perished when the French threw in their lot with Hitler. One perished on French soil (Münzenberg, hanged in the woods near Grenoble, Heinz Hirth was the prime suspect—KPD member, Comintern operative, and suspected OGPU agent). The other perished on German soil (Breitscheid, deported back to Germany, the Buchenwald concentration camp, where he was put to death).

The Nazis also rolled out the civil service law (April 1933), which became the vehicle to cleanse the government administration of undesirable elements, including the undesirable Jews, as not all were undesirable, at least not immediately. Eugen Mittwoch, head of the Oriental Languages Institute, where Taghi Erani had taught, carried on for some time, possibly his background in the intelligence service saved him, and Friedrich Rosen had retired, but nevertheless left Germany on the pretext of visiting his son in Beijing (the German Consul there, sacked later). And Lise Meitner continued her scientific work on induced radioactivity at Dahlem, where Erani's laboratory was also situated, though she was stripped of her teaching post. (She was Austrian, not yet subject to all the exclusion laws; that came when Germany annexed Austria.)

Taghi Erani knew these individuals personally or knew of their work, thus for him, these developments were not abstract, and soon he would visit Germany and see things for himself, autumn 1934 and summer 1935. Overall, the country went on a war footing, industrial recovery fueled by armaments production. Unemployment dropped, but this was the famous medicine being worse than the disease. War became imminent, a foregone conclusion.

Buttressing this was Schacht's concept of collective autarky, boosting barter with new markets (Latin America), minimizing exposure to countries with which Germany would be colliding. Persia was one such market, which saw exponential growth in trade volume, and unsavory Germans homesteading in Persia (including Wipert von Blücher, the ambassador whose invitation to Taghi Erani was for the debut of a Third Reich propaganda movie, *Deutschland erwacht—Ein Dokument von der Wiedergeburt Deutschlands*, in Persian *Bidariye Alman, Awakening of Germany*, screening at Mayak theater, Tehran).[12]

Despite these developments, there was a bizarre sense of complacency in major European capitals, London and Paris, even in America, not amongst artists and intellectuals who were quite sensitive and reacted with great energy, but amongst the political class, who seemed to think that this is what will sort out the Bolsheviks: another German assault, another Russian collapse, another Brest-Litovsk.[13]

But with time, the realization sank in that German knowhow, combined with Russian resources, would be an unbeatable combination, turning continental Europe into a fortress, what Napoleon had attempted with the ruse of republicanism sweeping moribund monarchies, cutting out the British from the continent, spelling the doom of its empire. This shift in mind-set, however, happened rather late, in late 1930s. Thus, Germany going after Russia became a "red line." This put an end to the reign of "guilty men" (title of a bestseller at the time in England), leading politicians who had slackened on rearmament, had flirted with Hitler (appeasement), turning a blind eye to his territorial acquisitions (Sarre, Czechoslovakia, Austria).

Persia's neighbor to the north was also experiencing serious convulsions in the late 1920s, early 1930s. Confident from eliminating the bulk of his opponents, and fearful that sooner or later the West would crush him, Stalin put Russia on a strict war diet. Termination of the NEP (New Economic Policy, which had allowed farmers to sell their produce on the market, as opposed to handing it over to the state at knockdown rate), roll out of the five-year plan (1928–1932), in the midst of which he gave his famous *"we catch up or they crush us"* speech (1931). This comprised forced collectivization of agriculture, extermination of the Kulaks (landed peasantry, some of whom fled to Persia), and relentless foreign policy imparted not only through official diplomatic channels but also through proxy conduits, Comintern and the intelligence apparatus in its successive incarnations (Cheka/OGPU/NKVD, pre-KGB).

In Afghanistan, the Soviets intervened militarily (eight hundred Red Army soldiers disguised as tribal Afghans, 1928) to save the monarchy of Amanullah Khan, facing a peasant movement that had resorted to radical land reform (*Bach'e-Saqqa* leading it, "water-boy," referring to his modest origins). The Afghan king's power base in the south ensured he would adopt a tough stance there, bordering India, stonewalling the British, but would be lenient toward the Russians in the north (Turkmenistan, Soviet territory). It was interests not ideology— Bolsheviks backing a feudal monarchy in the face of revolutionary peasantry—that was the guiding principle of Soviet foreign policy from the very beginning.[14]

This mode of operation originated from Lenin, as exemplified in how he handled the Persian question, a bargaining chip with the British, leading to the Persian–Russian and Anglo-Russian agreements (February/ March 1921). Lenin played with insurgencies in Gilan and Khorasan, blowing hot and cold, through his trusted intermediaries (Rothstein, Karakhan, Litvinov, Chicherin). All the posturing about internationalism and world revolution and the right of nations for self-determination came to an end once the Bolsheviks seized power, with a period of ambiguity during the civil war, though the rhetoric continued for quite some time. (The forgiveness of the debts was also a way of not paying back their own debts, far more onerous, particularly to the French.)

In Persia, the changes were not any less momentous, by its own standards: the 1929 oil strike, the first assertion of working-class power, the conjuring notions of Red Persia; what General Ironside had foretold in 1920 (the people destitute, the elite rotten, the country ripe for communism). And another English commentator, Reginald Bridgeman putting it more poignantly in 1931—had it not been for Persian oil, the British government would have collapsed in the face of the general strike (May 1926, tanks in the streets of London, nine hellish days, one and half million strikers, miners and railway men at the core, and hired strikebreakers). The loss of Persia and the APOC was an existential threat to Great Britain.[15]

Hence, the military crackdown against the 1929 APOC strike (walkout of nine thousand workers at Abadan refinery, rocking other oil towns), which had set forth demands not only for improved wages and working conditions but also the right to unionize, worker's clubs, and the recognition of May Day. The response to this was the deployment of a British warship on the nearby waterway (Shatt-al-Arab or Arvand River, a cruiser armed with six-inch cannons). Even provincial police mobilized a hundred automobiles in one of its crackdown operations, scale of mobilization unheard of at the time.[16]

Then came a godsend, Georges Agabekov, the OGPU chief in Tehran, defecting to the West, 1930, divulging what he knew about Soviet operations in Persia and their native collaborators, with a far greater precision than Bajanov could have ever done.[17] Bajanov had defected in 1928, crossing the Turkmenistan border on New Year's Eve while border guards were dead drunk, and he managed to escape the Soviet network in Persia. Then, with the help of Persian officials, he crossed the border to India, and from there sailed to London. His information was mostly about the Kremlin, as he had been one of Stalin's secretaries for a few years.

The APOC strike, which agitated several cities in the south (Abadan, Ahvaz, Masjed-Soleiman), and other strikes which came in its wake (*Vatan* textile in Isfahan, *Kebrit'sazi* or matchsticks manufacturer in Tabriz), and the Agabekov defection collectively led to massive arrests of labor and political activists, former organizers and leaders of trade unions (banned since mid-1920s), veteran communists and left-wing sympathizers, as well those deemed close to Soviet circles (tutors, translators, drivers, and so forth). Two hundred were rounded up, four sentenced to death, at least one carried out.[18]

A legislation was pushed through *Majles* with lightning speed, the Court Minister and Justice Minister, Teymurtash and Davar, working around the clock to calm the Shah's nerves, with the penalty of three to ten years of solitary confinement (*habs'e mojjarrad*) for anyone engaged in anti-government activity. They were asking the public to alert the authorities of suspicious individuals and activities, while offering amnesty to those who would step forth and denounce their co-conspirators.[19]

In the midst of this came the negotiations for renewal of the oil concession, having come to its thirty-year term in 1931. The entanglement of the Court Minister in this process, his flirtation with Moscow and alleged linkage highlighted to the king by Persia's intelligence service, and presumably by British intelligence based on photographs of documents to which they had gained access, all boiling over in the summer of 1932. (Teymurtash had led the trade negotiations with Russia, concluded in 1927, and in 1932 had hinted about Russian involvement in Persian oil, having been received at the Kremlin on his return journey from Europe; he was perfectly fluent in Russian, a connoisseur of its literature, having boarded there for eleven years in his youth, and was on first-name basis with Litvinov and Karakhan, the Commissar of Foreign Affairs and his deputy. Stalin was also present at the reception held in his honor.)[20]

Then came Teymurtash's abrupt dismissal (December 1932) and an article in the *Times* of London (January 1933) that in British eyes he had become even more powerful than the Shah, virtually a death sentence when brought to the attention of the insecure monarch, causing a legal case to be put together, charging him with rigging the exchange market with the help of Persia's central bank president, Herr Kurt Lindenblatt. This charge was true, but not the real motive of the arrest, as the Shah was a co-beneficiary. Lindenblatt had to be punished as well, since his assistant, Herr Volger, had lost great sums of money in financial speculations with royal funds, sourced from oil revenues. Volger was a German national but carried a Bulgarian passport, fled Persia for Beirut, was arrested there by French authorities, and presumably committed suicide. Lindenblatt did better, perhaps his connections saved him, or it was undesirable to draw attention to Persia on that score. He was dismissed in absentia (was out of the country at the time, hence his assistant was at the forefront), reached a settlement and returned to pay his fines and submit to eighteen months house arrest.[21]

Teymurtash was eliminated soon after his sacking, at *Qasr* prison in the outskirts of Tehran (October 1933), his elimination oddly coinciding with Karakhan's official visit to Persia. Persia's intelligence service and its chief (M.-H. Ayrom) played a prominent role in the Teymurtash affair (detailed in his interview with the British consul, T. V. Brenan, March 1933).[22] This was a clear sign that the intelligence apparatus had gained the upper hand over the political class. There was no intermediary between the Shah and the Chief of Police, who was more powerful than any minister, and even the Prime Minister (position of Court Minister disappeared with Teymurtash). These, presaged a wicked winter at the pinnacle of power.

Taghi Erani observed these developments from close quarters as he knew many of the actors and victims. Given that 1933 was the final year of the "quiet period," as from then on, he went on a path of no return, it is clear that it was a year of intense introspection and inner struggle. Was it the fugitive fate of those he had known in Berlin (Münzenberg, Breitscheid, Rosen in some sense)? Or was it the rise of the Third Reich and homegrown fascism; the decadence of the domestic elite, the misery of the masses, the morale of the youth, the militarism of European powers, and the coming of the Great War part two which would drag Persia into it, as it had the previous time? Whatever it was, it prompted him into action.

He had kept a low profile long enough to be cleared of any suspicions, so he thought anyway. And he had done sufficient groundwork over these years, both intellectually and logistically, to be able to work out his game plan for the next few years at least. But what did he contemplate and how would he go about it? His deeds the only indicator of the realness of whatever we may surmise in terms of his reflections, thought process, mental universe.

Notes

1. See Momeni, *Parvandeh*, pp. 227–228, 256, for Erani's assignments and rank at the Ministry of Education. See Boroujerdi, *Erani Faratar az Marx*, pp. 507, 514, for government data on what he owed due to student financial aid. Dr. Mehdi Azar indicated that Taghi Erani opted not to join government service in official capacity, due to intolerance for bureaucracy; private communication.
2. This was Iraj Eskandari; see Dehbashi, *Khaterat'e Iraj Eskandari*, pp. 37–41.
3. Erani taught in the following schools—Iranschahr, Dar-ol-Moalemin (Teacher's College), Servat, Scharaf, Maaref, Ghoor-Khaneh, and Nezam high schools. He also taught at the newly formed University of Tehran. These are based on accounts of Khame'i, *Panjah Nafar va Se Nafar*, pp. 89–90; Momeni, *Parvandeh*, pp. 227–228, 256; Boroujerdi, *Erani Faratar az Marx*, pp. 615, 627; and Ayrom (personal communication).
4. Mr. Ayrom, no forename given, private communication. His letter came about due to an ad Dr. Mohammad Assemi put in the *Kaveh* monthly (Munich-based journal, 1970s–1980s), to help the author seek information on the life of Dr. Taghi Erani.
5. See Momeni, *Parvandeh*, pp. 89, 294–297, for list of books taken away by the police in multiple searches; see p. 246 for description of Marxist books in English that he bought in London, with mention of Brukheim bookshop in Tehran where he forwarded copies of *Capital*; journals he received from Europe are mentioned on pp. 71, 88.
6. See Khame'i, *Panjah Nafar va Se Nafar*, p. 14, for physical attributes and movement habits of Taghi Erani; see Erani, *Osool'e Elm'e Rooh*, pp. 142–143, on mental processes causing the subject to be unaware of details of movement and itinerary.
7. Trip to Mazandaran, described in police interrogations, see Momeni, *Parvandeh*, pp. 57, 76. Trip to Varamin noted in photograph annotation, see Chaqueri, *Taghi Erani dar Ayeneye Tarikh*, p. 203. Trip to Tajrish and mule story from Mr. Parviz Shokat's inspection of Erani manuscripts

(private communication). In Erani's writings (*Psychology*), we find names of regional oils and distillations, showing familiarity with domestic geography.

8. See Neukratz, *Barricades of Berlin*, for account of May Day 1929, banned by social democratic government (Hermann Müller), thirty-three workers were shot by the police; Wedding and Neukölln described in detail. See Haffner, *Failure of a Revolution*, p. 142, for German revolution of 1918–1919, and Liebknecht's eight-word speech against the war given in May 1916. For Erani's observations of working-class Berlin, see Erani, *Donya*, 12-issue Volume, p. 329 (in *Bashar az Nazar'e Maadi*).

9. Comintern's conditions of membership for political parties were drafted in 1920 based on Lenin's formulation; condition sixteen stated that all decisions of Comintern Congress and Executive Committee are strictly binding on the member parties. Certain parties split because of this condition, one faction joining the other not (e.g., the French Communist Party).

10. See Hallas, *The Comintern*, pp. 91–102, for an account of the "German October" of 1923.

11. See Stoltzenberg, *Fritz Haber*, pp. 164–165, for an account of the Rapallo Treaty.

12. For Schacht's vision as it concerned Persia, see Jenkins, *Hjalmar Schacht, Reza Shah, and Germany's Presence in Iran*. For German-Persian trade relationships during interwar years, see account in Valizadeh, *Etehad'e Jamahir'e Shoravi va Reza Shah*, pp. 243–260. Wipert von Blücher invitation is in the *Erani Family Archives*.

13. See Freund, *International Congress for the Defense of Culture*, for an account of massive congregation of intellectuals against war and fascism in Paris (1935); complacency in Europe came in the form of appeasement (Chamberlain); in America, it came in the form of whitewashing Hitler in influential media (*New York Times, Chicago Tribune, Christian Science Monitor*); see Reiss, *The Orientalist*, p. 283.

14. See Agabekov, *O.G.P.U.*, pp. 158–170.

15. See Ironside, *High Road to Command*, p. 153; see Bridgeman, *Persia and British Labour*.

16. See *Peykar* account reproduced in Najmi Alavi, *Sargozasht'e Morteza Alavi*, pp. 59–60 (hundred automobiles mobilized); also alluded to in Bridgeman, *Persia and British Labour*. See Cronin, *Popular Politics*, for detailed account of the 1929 APOC strike.

17. See Agabekov, *O.G.P.U.*, p. 178—"Persia thus became the center of espionage into India and Irak; Teheran became a second Berlin, for Berlin, you know, is the OGPU center for Europe."

18. See Valizadeh, *Etehad'e Jamahir'e Shoravi va Reza Shah*, pp. 66–67.

19. See account in *Peykar*, reproduced in Najmi Alavi, *Sargozasht'e Morteza Alavi*, pp. 45–49; also see Erani, *Defaiyat*, he calls it the "Black Law" (*ghanoon'e siah*).

20. See Sheikholeslami, *So'ud va Soghoot'e Teymurtash*, pp. 307–311, for Ayrom's role in this affair. See Pasyan and Motazed, *Zendegi'e Reza Shah*, pp. 496–497, memoires of governor of Gilan (Reza-Ali Divan-Beygi), who received *Tass* reports of Teymurtash's stopover in Moscow; he alludes to the theft of Teymurtash papers by British intelligence and disclosure of its contents to Reza Shah.

21. See Sheikholeslami, *So'ud va Soghoot'e Teymurtash*, pp. 292–297, for the *Times* article. See Motevali, *Teymurtash va Baziy'e Ghodrat*, pp. 101–104, for ending of Volger and Lindenblatt. See *Peykar* account, in Najmi Alavi, *Sargozasht'e Morteza Alavi*, pp. 146–148.

22. See Sheikholeslami, *So'ud va Soghoot'e Teymurtash*, pp. 307–311, for memo of T. V. Brenan (British consul in Tehran) to R. H. Hoare (British Ambassador to Persia) concerning his interview with Mohammad-Hosayn Ayrom, head of Persia's intelligence, in regard to the Teymurtash affair (FO 371/16951/9563). See Pasyan and Motazed, *Zendegi'e Reza Shah*, pp. 551–557, on Karakhan's visit to Persia and Ayrom's manipulation attempting to have him visit the *Qasr* prison, and elimination of Teymurtash at precisely that time.

CHAPTER 13

The Post-quiet Period (1934–1937)

The quiet period (1929–1933) was quiet only in so far as there was no explicit political action on the part of Taghi Erani. His actions seemed to revolve around academic, professional, personal, and social preoccupations. But appearances can be deceiving, and from documents to which we now have access, we know of five contacts that were made during this period with a definite political purpose and intent. These leave no doubt that the quiet period was quiet only in appearance; the second part of this period was nothing less than preparing the ground for the next phase of action.

The first contact was with an acquaintance who had lived in the same neighborhood in Tehran when they had arrived there from Tabriz (1912). Their landing spot was the little Bazaar (*Bazaarche*) at *Sheikh Hadi*. Their neighbor was not from Tehran either, but a small town some 200-km away (Arak). Their son was a *Dar-ol-Fonun* student, a little prodigy named Abdol-Hossein Dehzad (Hesabi), but when Taghi Erani started at *Dar-ol-Fonun*, he had already graduated (b. 1895, graduated 1911). After graduation, he attended the School of Political Science (founded in 1899 by an enlightened Qajar aristocrat, Moshir-ol-Doleh, a two-year curriculum), worked as a clerk in the Ministry of Finance, part-timed as a high school teacher, and dabbled in journalism, satire, poetry, and political activism.[1]

Dehzad was locked up for a brief period (protest against the Anglo-Persian Agreement), and upon his release in late 1919, he dropped out of sight, joined the insurgency in Gilan, enlisting with the left-wing

© The Author(s) 2019
Y. Jalali, *Taghi Erani, a Polymath in Interwar Berlin*,
https://doi.org/10.1007/978-3-319-97837-6_13

forces of this movement (Ehsanollah Khan), who seized political power in the province for a year (1920–1921). He became editor of this movement's flagship journal in Rasht, *Lava'ye Sorkh* (Red Banner).

When the movement was crushed in the summer of 1921, he went to the Soviet Union, trained as a party worker at a school the Bolsheviks had just set up in Moscow for the "peoples of the East" (KUTV where Ho Chi Minh, Deng Xiao Ping and many others were trained), then returned to Persia in disguise, and engaged in a teaching career. He was arrested after the coronation of Reza Shah (April 1926), possibly with a different identity, and was released after several months. He left Persia a second time.

When the Comintern decided to activate the network of Persia's communists, they chaperoned its party congress on Russian soil (in Rostov or in Ivanova, both have been cited, October 1927) with a program that was conducted mostly in Russian; a congress to which Stalin, Bukharin, Thälmann, and other notables were elected as honorary members.[2] This gathering was, however, advertised as having taken place in Persia (Urumi'ye, Azerbaijan), as the party's second congress (though it was its third, the first held in 1920 in Port Enzeli, second in 1921 in Baku, a postmortem critical of Bolsheviks and native communists after the collapse of the Gilan insurgency, a gathering invalidated by the Comintern, thus the third congress was branded as the second). Dehzad attended this "second" congress, addressing the potential of trade unionism for revolutionary action (banned after Reza Shah came to power, but with a history of over two decades and notable successes, a print-setters' union, and an eight-hour workday). Soon thereafter he slipped back into the country (1928). This time he got engaged in industrial outfits, in Isfahan. This was a clear sign the Comintern eyed the working class in Persia as one of its levers. He seemed to have played a role in the strike action at *Vatan* textiles (1930). Dehzad was one of the few leading activists who went unnoticed in the police crackdown, in the aftermath of the APOC strike (1929) and the Agabekov defection (1930). But his position was precarious. He moved to Tehran, went incognito there for a year, and then crossed the border to Soviet Azerbaijan (Astara), and from there to Moscow, leaving behind his wife and son in Tehran.

While in Tehran, in late 1931, Dehzad made contact with Taghi Erani and met him several times in the outskirts of the city (*Golab Dareh*, Rosewater Valley).[3] The converse was not possible, Taghi Erani contacting him, as he worked with a false identity with unknown whereabouts.

He was fluent in French and Russian and had an intellectual bent. His information about Taghi Erani came through his hierarchy in the Comintern, and it so happened they had been neighbors in their youth for seven years.

The Comintern knew about Taghi Erani from Berlin, where he and four others founded the CRRP (1926) and solicited collaboration via Louis Gibarti, Comintern's liaison in Berlin. The Comintern was impressed by the logistical resources of this group, self-formed, clandestine, with access to Persian printing facilities. For a brief period, the Comintern considered infusing its dormant network in Gilan (via Ehsanollah Khan's connections, who was in exile in Baku) with the CRRP political platform and resources in Berlin, but dropped the idea, as they found the CRRP rhetoric not militant enough (compared to the rhetoric forged in the sixth congress of the Comintern, 1928, at the height of Anglo-Soviet hostility). Comintern then managed to put together Persian printing facilities in Berlin, with the help of the German Communist Party (KPD).

At any rate, Berlin was the hub of Comintern activity in Europe (Western European Bureau), and it is there that Taghi Erani came to their attention. Until the end of 1928, they could track him in Berlin; thereafter, their information was most likely provided by the CRRP co-founder, M. Alavi, who stayed behind in Berlin and maintained his connection with Taghi Erani.

Taghi Erani posed a number of questions to Dehzad about the specifics of the Gilan insurgency and subsequent party work, but he does not seem to have received explicit answers.[4] He was, however, told that someone would be in touch with him as soon as feasible.

The second contact was with the collaborator of his childhood neighbor. He was a bit younger than Dehzad, but with a similar background; an intellectual engaged in the practical end of underground movement, participant in the Gilan insurgency, a native of Mazandaran, the adjacent province (also bordering the Caspian). His name was Ladbon Esfandiari.[5]

He had had an education in French and was the younger brother of a well-known poet, Nima Yushij, the Baudelaire of Persia, the father of its modern poetry, blank verse—*as we stroll by the shore, insouciant, it claims yet another life, the deep dark sea...* Ladbon also dabbled in poetry, though his predilection was politics.

With Taghi Erani, he shared political literature put out in Vienna by Persia's communists (*Setareh Sorkh*, Red Star, theoretical journal, 1929–1931), and another put out in Berlin and later Vienna by what seemed to be a coalition of left-wingers, serving as a platform for anti-Pahlavi, anti-British political agitation (*Peykar*, Combat, 1931–1932).

Ladbon also shared the manuscript of a work he had completed containing an analysis of the causes of the 1929 world economic collapse (he has several works, though only titles seem to be extant). Taghi Erani shared his work on psychology, markedly his latest addition on sociohistorical and cultural psychology, in which he had dedicated a section on Persia's intellectual development from antiquity to the present.

This touched on pre-Zoroastrian nature worship, Mitraism, the emergence of Zoroastrianism corresponding to a sedentary culture, the development of social differentiation, classes, castes, slaves (war prisoners), and the crisis of this order as manifested in the failed Zoroastrian reform movements of Mauni and Mazdak. Then came the emergence of Islam as an ambiguous mix of invasion and social emancipation (of slaves), and post-invasion consolidation of an aristocratic order (Umayyad, Abbasid). The crisis in Islam was epitomized by a long sequence of political challenges to this oppressive order, defeat of rebellions and intellectual backlash in variety of introvert philosophy, ending with the emergence of Babism (Lutheran in spirit, dispensation with clergy) and its collapse in the nineteenth century.[6]

This broad-brush account ended with an overview of the last three decades, the onset marked by the assassination of Nasser-ol-Din Shah (1896), portending the development of a post-feudal society as epitomized by the constitutional revolution (1906), suffering a setback but with definite signs of it piercing through (nationalist insurgencies, 1920–1921). These he portrayed as being analogous to the Enlightenment preceding political upheavals in Europe, shifting mind-set as harbinger of change in political order.

At their final meeting, Taghi Erani passed on a copy of his treatise on psychology to Ladbon, shortly before it appeared in print in Tehran (May 1932). Ladbon left Persia, presumably through the same channel as Dehzad (the two regrouped) and wrote a letter to Taghi Erani (letter extant) from the coastal town of Astara, on the western edge of the Caspian Sea, in which he had both praise and critique of Erani's work.[7]

He found the work gave to psychology too much credence in motivating human action, as opposed to man's material and physical environment. The gist of his praise, somewhat back-handed, was that notwithstanding the work's bias towards the importance of mental life in shaping human behavior, it is marked by so much erudition that it is certain that there would not be a single soul in all of Persia capable of comprehending it. And this conviction, he reassured Taghi Erani, stemmed from his wide experience with people from all walks of life in Persian society. He counseled Taghi Erani to carry on with his academic endeavors and make his contribution to the intellectual rejuvenation of the nation, and not concern himself with political activism.

As a result, the two contacts, with veterans of political activism, came to an anticlimactic end, at least for the time being. Taghi Erani seems to have attributed this ending to the emotional state of the commentators (as the two seemed to share similar sentiments), rather than the limitation of his philosophy and outlook. Therefore, he decided to reorient his focus, turning his attention towards his Berlin connections.

The third contact was with a friend who returned from Berlin in 1931, who was somewhat bourgeois in comportment. His name was Morteza Yazdi. He was a medical doctor married to a German woman, who was an advocate of the National-Socialists (Nazis). He founded a society to which a select group of medical doctors subscribed and functioned as a mechanism for professional networking and advancement. This stood in sharp contrast to a second society formed several years later by another Berlin-educated medical doctor (Mohammad Bahrami), which inscribed in its charter contribution to social well-being as its sole purpose and ethos.[8]

Nevertheless, this acquaintance brought a package for Taghi Erani from Berlin, given to him by a common friend (Morteza Alavi, one of the co-founders of the CRRP), although he had nothing to do with political affairs. Morteza Yazdi simply had been a member of *Iran Society*, like all other Persian students in Berlin at the time, and most likely he was not aware of the contents of the package, which included a codebook so Taghi Erani could correspond with his former collaborator in cipher.[9]

Morteza Alavi had been able to stay behind in Berlin after the departure of Taghi Erani. He was registered as a student of economics at the University of Berlin and began collaborating with prominent Persian communists who were dispatched there from Moscow, after the Agabekov defection and breakup of the network in Persia. The crackdown meant that it was better to set up publications in Europe and smuggle it into the country as European parcels. They may also have wanted to leverage the German economic foray into Persia to have a supplemental axis of infiltration via native technocrats, and they maintained a contingent of Persian operatives in the Soviet Union.

But German police caught up with this scheme. Morteza Alavi had a female friend who was an undercover agent. Persia's embassy also filed legal charges against him for making defamatory statements against the Shah in his political journalism (which was anonymous). Within twenty-four hours of this filing, he was expelled from German territory. A court case, however, was held in absentia and amazingly he prevailed.[10]

The defense did not deny that the journal articles were by Morteza Alavi but held that these had appeared before the passage of the bill in Persia's *Majles* penalizing such act. The plaintiff (Persia's embassy) rebutted that this person has also made defamatory statements about the President of the German Republic. This too was dismissed, as he had bashed at the "German Sultan," and linguistic experts brought to the court maintained that it is not possible to prove that Sultan and President are synonymous. The defendant seems to have had a better legal team than the plaintiff.

Accordingly, he won his court case, in absentia, as he had settled in Vienna, and had resumed his journalistic work there. These events accelerated his rapprochement with Persia's communists operating in the Comintern, and indeed it was the KPD, a member party of the Comintern, that had given him all the logistical support, both for the move to Vienna (he transited through Prague), and the court case. Despite this legal victory, he did not return to Germany, and when Austria followed the path of Germany, after the rise of Hitler to power, he was dispatched to Moscow (1933). This, however, did not disrupt his link with Taghi Erani; even a year later, their communication channel functioned in both ways.

The fourth contact was with the younger brother of Morteza Alavi. They came from a prominent family, certainly on the paternal side. Their great grandfather had been treasurer to two Qajar monarchs. Their grandfather was a well-known currency-broker (*sarraf*) and deputy at *Majles*, which had a class-based quota system, favoring the gentry and the nobility (*ayan va ashraf*). Their uncle had been sent to Germany to study chemistry, but perished there after an accident in the laboratory, an explosion which claimed the lives of a number of students (Münster, c. 1908).

Their father was also sent to Germany before the war, to explore commercial opportunities for trade. But he was not cut out to be a merchant
and was more inclined toward political activism that would bring about
radical change in Persia, which even by Asian standards was backward.
The father collaborated with the pro-German party that operated out
of Berlin during the war (Persia Committee), while his family was back
home. He resumed his travels between Germany and Persia, toward the
end of the war and thereafter.

In late 1922, he brought along his two elder sons with him (he had
six children, three boys, three girls), arriving in Berlin in early 1923.
The eldest was Morteza Alavi (b. 1901), stayed in Berlin with the father
and decided to major in economics, and the younger was Bozorg Alavi
(b. 1904), moved to Munich and studied education and psychology. The
older brother and father got engaged in *Iran Society*; the older brother
got further engaged in the CRRP, and the father seems to have been a
sympathizer of this movement (his name does not appear in CRRP documents, but its chief manifesto, *Bayan'e Hagh, Attestation of Truth*,
is dedicated to his memory, using his pseudonym, after his suicide in a
Berlin train station, 1927).

The younger brother, Bozorg Alavi, no doubt moved by these events,
pursued his studies, and developed a passion for Western literature: short
stories, novellas, novels, plays, psychoanalytical fiction, and seems to have
developed a resolve to contribute to the creation of a similar universe of
immense possibilities and modes of expression in the Persian language,
stuck in the past, resting on its laurels, while the world moves on.[11]

Bozorg Alavi finished his studies in 1928 and returned to Persia, at
the urging of the older brother, to provide support to their mother and
siblings, who lived in the grandfather's home. He became a teacher of
German at a technical school in Shiraz. In 1929 he moved to Tehran,
again as a teacher of German, this time in the School of Industry (formerly German school), few months after Taghi Erani had returned
to Persia. By this time, he had begun his literary activities, translation
of plays (Schiller, Bernard Shaw), partial translation of a work on the
ancient history of Persia (Nöldeke), an essay on Nietzsche, and other
works he had in the pipeline.

Bozorg Alavi met up with Taghi Erani in 1930, and from then on,
the two developed a relationship of disciple and mentor, though he did
have another source of influence, on the literary side, from Hedayat, the
Kafka of Persia (*as I inhale my opium, comes to life images on the vase...*),

his older brother highly approving of the Erani connection. However, by his own admission, he was not cut out for political work; literature was his true passion, though there was something mesmerizing that drew him to Taghi Erani.[12]

The present author maintained a year-long correspondence with Bozorg Alavi, January–November 1986, seven letters, and also had occasion to meet him at the University of California, Berkeley, Dwinelle Hall, November 2, 1989. He passed away in Berlin in 1997.

The fifth contact was with a returning student from France who had studied law in Grenoble. His name was Iraj Eskandari.[13] He was younger than Taghi Erani, born in 1907, and came from an eminent family. His father had been a *Majles* deputy but had passed away in his son's childhood. His grandfather and uncle had looked after his education. His uncle was an enlightened aristocrat, *Majles* deputy, speaker of the minority faction (Democrats, second *Majles*), with ad hoc ministerial posts in late Qajar. His name was Soleiman-Mirza Eskandari.[14]

In the 1920s, the uncle had founded the Socialist Party of Persia, a name which in the original language (*ejtemai'yoon*) does not have the same connotation as it tends to have in European languages. It connotes communal and social endeavors, as opposed to socialist in an ideological sense. He was an advocate of meritocracy and egalitarianism in domestic policy, and autonomy in foreign policy. He maintained a positive outlook on the Soviet Union. He seems to have exerted a great deal of intellectual influence on his nephew, who had become exposed to the French left in Paris (1925–1927), before moving to Grenoble.

The Socialist Party of Persia was banned within a year after the coronation of Reza Shah (April 1926). The uncle journeyed to Europe in late 1927, with stops in Moscow, Berlin, Brussels, and Paris. In Moscow, he attended the tenth anniversary of the October revolution. In Berlin, he met with the CRRP co-founder (Morteza Alavi), whose late father he knew personally, and had kept abreast of their activities. In Brussels, he attended a second gathering of the League Against Imperialism. And in Paris he met up with notable socialists, accompanied by his nephew. He connected his nephew to the CRRP.[15]

A correspondence developed between the CRRP co-founder (Morteza Alavi) in Berlin and Iraj Eskandari in Grenoble. But Iraj Eskandari did not know the identity of the other CRRP principals.[16] When Court Minister Teymurtash went on his tour of Europe in 1928, he issued a tract in French (equivalent to the one issued in German, in Berlin) denouncing him as corrupt. He was suspected to be behind this act by Persia's embassy, so lost his government stipend. He carried on with family finances and returned to Persia in 1931.

Iraj Eskandari was debriefed at the Ministry of Justice (as he had been a student of law) about his political activities and was not only pardoned but given a post as an associate attorney.[17] This can be attributed to lapse of time from the alleged offense which was never proven, his pedigree, and also to the fact that the Minister of Justice was badly in need of technocrats familiar with the western legal systems (the Minister was Davar, Swiss educated, and had undertaken reform of the judicial system toward secular law). This resulted in his past being glossed over, and he thereby landed a decent job.

In 1933, Iraj Eskandari met with Taghi Erani in Tehran. As he was unaware of the identity of Taghi Erani in connection with CRRP, the most likely trigger for this contact was the CRRP co-founder in Berlin (Morteza Alavi), now in Vienna, soon Moscow. Intellectually, this fifth contact had a predilection toward jurisprudence, political economy, and political science. He seems to have viewed politics as a craft, more a profession than a passion. With Taghi Erani, they had intellectual exchanges revolving around the discipline of political economy, particularly Marx and his major work, *Capital;* but they also discussed political activity.

Iraj Eskandari produced the first translation of Marx's *Capital* into Persian, first two volumes published in 1970 and 1974 respectively, the third was begun but never published; in the preface to the first edition he indicated that this translation stemmed from a vow he had made to Taghi Erani back in the 1930s.[18]

***

Sometime in the latter part of 1933, Taghi Erani made a decision that would push him beyond the point of no return. One option was to do nothing dramatic, carry on with the status quo. This meant enduring the reign of Reza Shah, Persia's certain entanglement in the conflict brewing in Europe and Asia, disarray amongst the nationalist forces aggravating, and unfolding of a human tragedy of possibly genocidal proportions at home, as it had happened during the previous war (no less than one-tenth the population of Persia perished from the Great War and its aftermath—famine, epidemics, collapse of basic services, insecurity, banditry, hoarding).

Further, he may have reasoned that it is a law of nature, what befalls one is in some sense what one deserves; unless one is willing to take initiative and assume risk, a person has no right to complain of their plight. The Great War brought forth not only immense misery, but also a deep awakening, especially in Asia, particularly China and India, proving that the human spirit is not dead. Thus, unless one is happy living a life of pomp and frivolity, there is no choice but to take action. This predilection of Erani's was not stemming so much from cerebral considerations, but from his emotional temperament, a certain disposition flowing from his psychological makeup.

Taghi Erani had this type of predisposition, the cause of which is open to anyone's conjectures of a psychoanalytical nature—the averred inadequacy of the father, the selflessness of the mother, the role modeling with the siblings, the family's elite status, engendering a sense of patrician guilt, idealism, romanticism, heroism, and so forth.

But while the motivation to act may be emotional, the conceptualization of what to strive for need not be, and can be subtle and sophisticated. Its impact and durability would be the sole indicators of the soundness of the vision and the execution of the plan. Here, Taghi Erani had a formidable challenge. Had this been during the *belle époque*, before WWI, when the West was intoxicated by its progress and achievements, the formula could have been rather simple—imitate the West, and retain certain national characteristics to not suffer backlash from drastic rupture with the past. But all this was shattered after the Great War, and worse, the impending war made it impossible to pretend that the evil is not there, that one is only to follow the West unequivocally.

Hence, a narrative had to be forged which would suggest something more nuanced—that although the West is beset by crises, it can overcome it, as human history has shown all existential crises can be overcome (otherwise how did we arrive at our present condition, from base beginnings); the solution is not to go back, for clinging to one's antiquated traditions in a world that is moving on is nothing but retrogression, but to move forth, to have courage, imagination, resolve to innovate, to assail the difficulties and bring about a new existence, with fresh and inspiring perspectives, in essence leapfrogging.[19]

To make this narrative credible, it is imperative to take the long view, that in the long-cycle view of human history, one observes that when human groupings see within themselves the power to overcome their predicaments, they exhibit positive attitude, pushing aside doubt and trepidation, cloaked as wisdom and experience. Conversely, when the horde does not sense in itself the capacity to overcome forces menacing it, it sinks to negativity and rationalization of why it cannot be any other way, fatalism.

Today, this is seen not only in the East but also in the West and is not confined to the sphere of humanities, but has also spread to the sphere of science, its eminent representatives casting a shadow of doubt on the law of causality, existence of the external world, and its knowability, postulating instead that the universe is finite, and the certainty that it is beyond our grasp, chaotic and random. This is nothing but a deep crisis of confidence, a psychological meltdown.[20]

Once the causal relation between societal milieu and social psyche becomes clear, demonstrable with historical evidence, it can be grasped by those with greatest aptitude for rational thinking, instilling in those individuals a healthy acumen which spreads according to the law of contagion.

The depiction of this cognitive and emotional ebb and flow, stemming from man's material and mental culture and circumstances, across geographies and through the ages, is therefore a task of fundamental importance. This is what Taghi Erani set out to sketch in the fifth element of his *Series in Exact Sciences*, dedicated to "General Principles"—history of thought and logic. Specifically, one of his essays, *Mysticism and Materialism (Erfan va Osool'e Maadi)*, is focused on this theme. Recognizing where one's culture falls within this landscape is of great importance, guiding one's field of action, which is always local.[21]

In undertaking this task, one arrives at a concept which can be called *monism*—that there is no immutable, unbridgeable gap between categories with which our thought processes tend to get afflicted, for all such categories are ephemeral and transient—science and philosophy, philosophy and religion, religion and mythology, mythology and magic, all is a continual process of cognition, one form subsumed into the other, causing its transmutation, an anti-essentialist viewpoint.

The same holds for other prominent dichotomies and cleavages—natural and human sciences, East and West, past and present, savage and civilized, backward and advanced, weak and strong, man and woman, and so forth. This logic, of transformation, interaction, reciprocation, is manifested in the tension between immediate and historical causation, continuity and rupture, and captures natural and human phenomena in its richness and complexity, enabling us to grasp reality to a great degree of authenticity.

This concept is so pronounced in the mental universe of Taghi Erani that he gives its designation in three languages (French, German, Persian)—*dialectique universelle, Dialektik Universal, Asl'e Takapu*. Underpinning this logic is the philosophy of materialism; that all things flow from other things, that there are no phenomena not preceded by other phenomena, as there is no phenomenon not engendering other phenomena, and that the material world in its most general and all-embracing sense (matter-energy in the unbounded manifold of space-time) exists independent of the observer.[22]

This conceptual framework, according to Taghi Erani, is what liberates us from the shackles of recipes and formulas, trumping all creeds ending with "isms," their dictums and dogmas, postulates and paradigms, allowing us to analyze any case on its own merit, engage with it, refine the perception on the basis of the outcome of the engagement, and devise plans to steer it toward desired ends, which can be influenced but not predetermined, as posterity is contingent and complex.

Thus, from an epistemological point of view, it seems that he felt he was on solid ground, despite the negative vibes he received from veterans of Persia's political activism. He had now to give to this vision the level of concreteness that it required to be a firm basis for action. But precisely, what type of action?

If his intent were to remain within the realm of cultural and intellectual activism, he certainly did not stick to that plan, and within nine months (on the first return trip to Berlin) veered toward political activism, the decisive turn taking place after thirteen months (engagement with a Comintern operative, alias Kamran). On the other hand, if he did conceive this step as a prelude to political activism, it is not clear what plan if any he had for this subsequent phase. Our conjecture is that his intent was something more than cultural activism, yet he was unclear as to how to proceed from the cultural to the political sphere; most likely this is something he imagined he could work out down the line. That he did, but it turned out to be a turn into the abyss.

As fuzzy as this game plan was, certain things were clear, according to his view of things—seizure of political power is the precondition to implementation of a program of social rehabilitation, national rejuvenation, and modernization. Political power is not something that can be attained by consent or peaceful means, at the ballot box or in the parliament; even the legal historian Rudolf von Jhering, bourgeois in outlook, attests to this. What is required is a revolution, a radical move to grab political power, and what makes for a successful revolution is a political party capable of leading it, with a clear vision of how to reinvigorate the nation, hold on to power without degenerating into an instrument of autocracy.

The timing for such revolution cannot be predetermined, nor its precise modality (Russian model, Chinese model, Indian or Indonesian model, all models are useful, though ultimately it has to be a native model), but the party can work towards its realization and possibly trigger its outbreak. What is required to have such a party is a committed group of professional revolutionaries. Persia's revolutionaries, however, were either locked up in dungeons or dispersed in exile and possibly sapped of their vitality, feeling demoralized.

One could play for time until those in jail won their freedom or developed a more positive outlook, but time was of the essence, the drumroll of war on a crescendo and the local elite playing both sides of the war game; this guaranteed that one way or the other this scourge will besiege everyone. Thus, a new generation of professional revolutionaries has to be brought forth, waiting was futile.

The stifling cultural ambience of Persia could conceivably be stirred through a journal guided by a definite philosophical outlook, to observe what groupings gravitate towards it and with whom this intellectual renaissance shall resonate. Then, from amongst those stirred into action, candidates can be spotted for disciplined organization. This can span the capital and the provinces, embracing people from all walks of life, the youth, intelligentsia and working classes—students, teachers, laboring types, clerks, technocrats, and so forth.[23]

To reach as wide an audience as possible, such a journal must operate within the existing legal and regulatory frameworks; hence, it could not be explicitly political. Its thrust could be philosophy, social sciences, and political economy (stressing how modern civilization is founded on the basis of modern modes of production, boosted by trade and mercantilism and judicious distribution of wealth, all contingent on the abolition of antiquated modes of production to liberate the laboring forces, notably the peasantry). The journal could contain data and analysis on economic matters (drawing from major European sources on the post-depression landscape in Europe, America, Japan, their civilian and military expenditure, balance of trade, unemployment). The journal could also include cultural commentary (art, literature, book trade, feminism), world events (chronicling the march toward war and anti-war pacifism in Europe, note that Wipert von Blücher's invitation to Taghi Erani to view the propaganda movie *German Awakening* came at precisely this time), news of scientific and technological innovations, and the like.

Such a journal needs collaborators to get off the ground and be credible, the more mainstream the better. Additionally, it needs financing, which can be partially secured by commercial ads from Persia's thin but thriving class of technocrats (doctors, lawyers, industrial firms), adding another layer of dissimulation to the undertaking. It also needs grassroots support at the level of publishing staff and print-setters to circumvent draconian rules of censorship. Thus, was derived, as if by mathematical deduction, the plan for the genesis of an *avant-garde* journal in the height of political and cultural repression and blackout in Pahlavi Persia.

***

In January 1934 (Bahman 1312, in the Persian calendar), there appeared in several bookshops in Tehran (Brukheim, Avicenna, and one named Tehran), the first issue of a journal with a green cover, with an image of the atom-smashing experiment of Cockcroft and Walton (vapor traces), captioned as "*Man Disintegrates the Indivisible*"; thirty-two pages in length, smallish cut (B5 format), dedicated to topics in science, industry, society, and the arts expounded from the perspective of materialist philosophy. Its name suggestive of its scope—*Donya* (the world), editor-in-chief Dr. Erani, publisher *Ettelaat*, address: Ministry of War Junction, *Ettelaat* Building, *Donya* Bureau.[24]

The journal would feature a leader (first column), essays relevant to its scope and slant, as well as a column on world events (*Manzareye Donya*), recent publications in European languages, tidbits on general knowledge, and a Q&A with the readers. It would also contain ads—medical clinic, pharmaceutical products, dictionary, monograph, novel, highlighting author and publisher, and publications by Dr. Erani (*Series in Exact Sciences*, notably *Psychology*, and *Khayyam on Euclid*; the latter appeared with fifteen-month delay from announcement). Its periodicity would be monthly, an annual subscription cost of two *Tumans* (equivalent to two dollars, or ten Marks), and a single issue a tenth of this rate (two *Krans* or *Rials*). This journal could also be purchased through mail, by sending in a two-*Rial* stamp. The production volume was about two hundred copies per issue, presumably half for the capital, half for major cities in the provinces (certainly in Shiraz and Abadan it had distributors).[25]

The next five issues came out on a regular monthly basis (issue four onwards published by *Sirousse Press*): cover photographs including—a motion picture recording device, giant communication antennas, a giant airplane (twelve engines), an airship called USS Los Angeles (made in Germany), and a photograph of Madame Curie and her daughter Irène ("*Women at the Cutting-Edge of Science*").

Then came issue seven with a one-month delay (Klingenberg power plant, Berlin, cover photograph), issue eight with a two-month delay (Eiffel tower cover), and issue nine with no delay (Irène Curie, discoverer of induced radioactivity, cover). Then came combined issues ten, eleven, twelve with a nine-month delay (September 1935), though nominally a five-month delay. This last issue had the Doble steam bus on its cover photograph, with the following commentary—it is fueled by kerosene, more cost-effective than gasoline, the innovation stemming from scarcity of petroleum resources, a major cause of global tension and conflict. There was also an announcement that the new series of *Donya* would appear from the start of the following calendar year (March 1936), targeted for a wider readership. This, however, never materialized. The twelve-issue collection amounted to nearly four hundred printed pages.

The final issue, nearly one hundred pages, contained phrases at the bottom of most pages with statements of a philosophical and social character, in boldface—monism of nature and society, the fallacy of teleology, the determining role of human intervention in societal evolution, the arbitrariness and transience of divisions and cleavages in nature and human society, matter-energy, space-time, particle-wave, national–international, race, class, gender, tension between rights and privileges, the principle of progress through strife, monolithic society a myth, struggle of classes and factions universal, evolution-by-negation, scourge of fascism, chauvinism, illiteracy, renovation of present-day societal model as necessity to surmount prevailing global crises and secure world peace—the greatest imperative of our time.

Besides Taghi Erani, the journal had ten contributors. Six had a single contribution each, four of which were translations. One was from Taghi Erani's first sister, the translation of Helen Keller's *Three Days to See*, signed by Irandokht. Another was by D. Rajabi who translated a text on aviation technology (a two-part series). N. Behbahani translated a text on the design and construction of the Eiffel tower. A. Ghezel-Iyagh wrote on Morocco's champion of anti-colonialism, Abd el-Krim Riffi and his return home, extracted from a French journal. There were also two non-translation articles. A. Khame'i wrote on history of science (initialed A. Kh., was a former student of Taghi Erani, with whom the author corresponded, he inserted references to Erani's essays for elaboration of the article, thus was done with a good deal of Erani's coaching). Hossein Afshar wrote on mathematics of integrals, calculation of maxima and minima, two-part series (he was by one account a *Majles* deputy).[26]

The remaining four were more regular contributors. Two focused on science and technology, based on the free adaptation of European texts. A. Nur contributed two articles, one on measurement of speed of light using peculiarities of Jupiter's moons, another on the operating principle of the television. A. Ashtari contributed five articles, one a two-part series (wireless technology), as well articles on sound imaging, reading machines, automobiles, and television with natural colors.

A third focused on law and economics, the Grenoble-educated lawyer who did not use his name (Iraj Eskandari), but worked with a pseudonym (A. Jamshid), and as well contributed to the financing of the journal. He had five contributions, one a two-part series (money from the standpoint of economic theory), as well articles on the theory of value, objective origins of law, choice and necessity, and *Machinisme.*

And the fourth focused on art and literature, the aspiring novelist (Bozorg Alavi), Munich-educated (using the pseudonym F. Nakhoda, also contributing financially), four contributions (two on art, one on dream interpretation, one on feminism), and serialized translations and snippets on authors like—Stefan Zweig, Upton Sinclair, Somerset Maugham, Ernst Toller, Benson Y. Landis, Maurice Maeterlinck, Louis-Ferdinand Céline, and native artists—Hedayat (fiction), Nushin (theater), Minbashian (music), Darvish (painting).

While these ten contributed to roughly half of the contents of the journal, Taghi Erani contributed the other half (and the bulk of the financing), and he did this using three signatures—two named (his own name, and A. Ghazi, a pseudonym), and as well *Donya Bureau,* and certain articles were not signed, but done by him (Q&A section, knowledge tidbits, announcements). Several of his articles also appeared in the name of the other two pseudonyms (one on evolution signed F. Nakhoda, the other a subset of his elaborate article on dialectics signed A. Jamshid). One of his articles was an extract from his *Psychology,* though it had undergone editing. Another, on the theory of relativity, was extracted from his *Theories of Science (Teorihaye Elm).*

But the bulk of his writings were fresh contributions, appearing as a two-part or three-part series, with a strong emphasis on philosophical themes and history of ideas. *Mysticism and Materialism* (*Erfan va Osool'e Maadi*) traced the evolution of thought from antiquity to the present in the East and the West, touching on mythology, religion, philosophy and science, highlighting the lows and the highs of social mood and culture, as reflected in mystical movements and materialistic schools of thought, Bergson and proponents of positivism and skepticism the latest manifestation of low in the West. More epistemological was *Materialism and Dialectics* (*Materializm'e Dialektik*), and more cross-disciplinary was *Man and Materialism* or *Man from Materialist Viewpoint* (*Bashar az Nazar'e Maadi*), a hybrid of biology, sociology, and psychology. These essays collectively constituted the fifth and final element of his *Series in Exact Sciences* (*Selseleye Oloom'e Daghighe*), the preceding four being—physics, chemistry, biology (not extant, if written, though he produced three essays on evolution), and psychology. The fifth element was labeled "Logic and General Principles" for which he coined terms in German, French, and Persian (*Dialektik Universal, dialectique universelle, Asl'e Takapu*), and was the essence of his conception, a monistic philosophy, as alluded to earlier.

Now we need only follow the tread of events to understand what reaction this initiative stirred, what brought an end to the *Donya* project, and how this intellectual undertaking metamorphosed to political activism and the downfall of its mastermind. The gaps in the publication history of *Donya* give away nearly the whole story, when put in the context of information that came to light in the 1990s, revealed in *Parvandeh* ("The Dossier").

***

The lead article in the sixth issue of *Donya* (June 1934) was an assessment of the reaction the journal had stirred in the country. It was prefaced by a description of the sociological morphology of Persia, the countryside representing the bulk of the population (sedentary and nomadic), but cut off from main currents of modern culture, including the press.[27]

Among the urbanites, the middle and higher echelons had the greatest access to literature. Amongst these, certain individuals became fierce opponents of the journal and its guiding philosophy. This included not only enemies of secularism, on grounds of religious dogma and obscurantism, but also secular currents satisfied and satiated by the status quo. Amongst the latter, certain individuals resorted to hardball

tactics to sabotage the journal, by attempting to rob it of its technical capabilities (this was the most likely cause of switch from *Ettelaat*
to *Sirousse Press*, from issue four). Others resorted to slander, largely
stemming from national chauvinism, and how all we need is to resurrect
the monarchic grandeur of ancient Persia, a process claimed to be well
underway.

A second current, however, was critical of *Donya* but did not resort
to slander and raised thoughtful critiques in some journals. *Donya*, however, was convinced that if this group carefully reads and reflects on its
contents, certainly by the time the foundations of its thinking will have
been fully elaborated (stated to be a one-year program), they will gradually overcome their difficulties.

And finally, a third group was enthusiastic about the appearance of
this journal, extracts from a number of their letters were provided. They
saw the journal as a refreshing change in the stifling atmosphere of literary production and journalism in the country. Variations on this theme
were reflected in the tenor of the respondents. The article also indicated
that young educated women have expressed interest in *Donya*, and for
us they represent an important social group, an integral part of sober
intelligentsia, and we will have further contributions of interest to this
group. The caption of its cover photograph highlighted this attention
to women—"*Madame Curie proves to the world the cultivable talent and
potential of women.*" Thus, it seems that the first milestone Taghi Erani
had set for the journal, to shake up the stifling cultural climate, and spot
potential sympathizers, had been achieved.

Then, in the next four months (July–October) a single issue appeared
(no. seven, August 1934), and when issue eight came out (November
1934), there was a one-page announcement indicating that Dr. Erani
has just returned from Berlin with the manuscript of Khayyam's critique of Euclid, a collaboration with Dr. Rosen dating back to 1925,
which shall come to print shortly, by *Majles Press*, with addenda that has
recently been uncovered in Sepahsalar library in Tehran (Tusi's critique
of Khayyam, thirteenth century). The work was to be produced with the
assistance of local experts on ancient manuscripts and Arabic philology
(this publisher dropped out, possibly due to the same disruptive efforts
Erani had alluded to earlier, so it was put out by *Sirousse Press* with a
fifteen-month delay, robbing Rosen of seeing it in print; Rosen passed
away in November 1935, the work came out in February 1936).

Therefore, during September–October 1934, Taghi Erani was in Berlin, and family photographs with friends in Berlin, in autumn clothing, substantiate this (as well as police interrogation records); the orientalist project and visit with friends, his only justification and cover for the trip. The outbound leg took place without going through the Soviet Union (possibly via Turkey).

In Berlin, a friend informed him that their common friend (Morteza Alavi, CRRP co-founder) had not stayed very long in Moscow (having arrived there from Vienna) and had moved to Tashkent. Erani wrote to him (most likely via the Vienna channel) to see whether the two could meet in Moscow, on his way back to Persia. He received a reply from Morteza Alavi saying that he will do his utmost to make it, but presently he is stranded in Tashkent. As it turned out, he did not make it to Moscow; Taghi Erani having arrived in Moscow proceeded on to Persia (via Baku), without seeing his former collaborator.[28]

The trip, however, was not a total loss. He did get a chance to see Berlin under the Third Reich, and how dramatically it had changed, since he had left in December 1928. Not only the leftist forces were persecuted and on the run, but also the civil service law had been put into effect. Even Rosen was not comfortable staying around. It is not clear whether he had already left for Beijing, or was in the process to do so. Therefore, we do not know whether the two met; the fact that Erani came back with their jointly worked manuscript suggests they may have met.

Eradication of *Stormtroopers* had also taken place (internal Nazi power struggle and bloody purge), as well as the book-burning charade, Freud, an Austrian Jew, a favorite target (about which Freud said—in the past they used to burn the author, now they burn only the books, what progress!). Overall, it was a chilling experience to see a degenerated landscape in Europe. There was therefore no such thing as over-reacting to news from Europe, it was daunting. The destructive machinery of war had been set in motion, and there were no prospects of it coming to a halt.

From then on things moved into high gear. In December 1934 came out issue nine of *Donya*. Then came an eight-month pause (January–August 1935), during which a great deal happened.

In February 1935, Taghi Erani received a note from the wife of his childhood neighbor (Dehzad), who lived in the outskirts of Tehran with her son in Ab-Ali, while her husband was in Moscow. She stated that a visitor who is in town is interested in meeting him. This was another native operative of Comintern, who knew the terrain well, as he had worked in industrial outfits in Isfahan and seemed to have a wide circle of connections. He had a professionally crafted identity, allowing him to move about freely. His name was Nosratollah Aslani, alias Kamran.[29]

Earlier, Taghi Erani had blocked off a contact by his former CRRP colleague in Berlin, Ahmad Asadi, the student of political science, the front man for the CRRP with links to the KPD; the first to be expelled from Germany (after his speech at the closed-door meeting denouncing the coronation of Reza Shah, April 1926). The one on whose behalf Taghi Erani made a plea to Rudolf Breitscheid. Asadi had gone to Moscow from Berlin, but was now in Persia without a changed identity. Taghi Erani found this odd, especially since he contacted Taghi Erani after the appearance of the first issue of *Donya*. Taghi Erani refrained from discussing politics, though they did meet several times. He pretended that those had been follies of youth, and he was content pursuing his personal and professional interests. This was a good reflex, as Asadi had become a turncoat, returning to Persia through the mediation of Teymurtash, Court Minister, and later turned into a Pahlavi regime diehard (he changed his surname and became a *Majles* deputy from Enzeli).[30]

Taghi Erani had been also approached by a lady, a certain Mrs. Kamali, who worked in a hospital as colleague of Morteza Yazdi, the bourgeois medical doctor from Berlin. She had political connections with a certain Shirinov (involved in setup of communist cells in Persia in late 1920s and early 1930s, independent of the Comintern, but at some point, began collaborating with the police). As her intentions were not clear to him, he deflected this approach. He also deflected another approach, by someone who had studied physics in France and was inclined toward political activism, with prior history of activism and arrest (Reza Radmanesh). But Taghi Erani had doubts about his mettle.[31]

But this time, a fourth time, the approach came through the spouse of his acquaintance of youth, in whose integrity and professionalism he had confidence, and their exchanges back in late 1931 had left him with a certain sense of an unfinished business. So, he agreed to the meeting with Kamran, which took place at a busy intersection in Tehran (Customs Square, *Meydan'e Gomrok*). They met several times, and the gist of this person's message was that after the police assault on the underground organization we must reconstruct the party, and you can contribute in several ways. One would be to steer your journal toward a platform aligned with organizational goals, the other would be to contribute financially to the reconstruction effort. Further, you could help by identifying candidates for party training at the school in Moscow for "peoples of the East."[32]

The first element was somewhat fuzzy. Did it mean to make the journal more political, clearly not feasible, or did it mean to convert it to a mouthpiece for the Comintern, abiding by their guidelines for overt journalism? Whatever the intent, Erani rejected the first element, citing infeasibility of politicizing the journal, and challenges concerning its sustainability, stemming from overstretched resources (contributors, time, finances), and administrative roadblocks. However, he did accept the second element and made a financial contribution (two hundred *Tumans*, equivalent of two hundred dollars, one thousand Marks), and agreed to think about candidates. In return, he was given an address in Moscow that could be used to contact Kamran, if necessary. He was also told that someone would be in touch with him, as soon as feasible.

Kamran moved about in the country and met with a number of people, including a medical doctor that had just returned from Berlin via Moscow (Mohammad Bahrami), a sort of people's man, and soon the founder of a medical association aiming to make contributions to social well-being. Kamran was escorted by Taghi Erani and this doctor and two friends, one a former roommate of Taghi Erani in Berlin (Ahmad Emami, he was also a medical doctor and had an automobile) and the other the math tutor of the king's male offspring (Mirza Abol-Ghassem Naraghi), to the Caspian area. They had been planning to go there for a short holiday for *Norooz*, the Persian New Year. The visitor tagged along and there were no discussions of a political nature. Such discussions would happen only on a one-to-one basis, and the two friends (Emami and Naraghi) were apolitical. Kamran was dropped off in Gorgan to meet up with his relatives, so he said. Most likely he took the ferry to the Soviet Union via Port Shah (southeast corner of Caspian Sea). His papers were in good order. This was in March 1935.[33]

In April 1935, the follow-up contact appeared. He was of similar age to Taghi Erani (b. 1902) and was perfectly fluent in Russian. His Russian was better than his Persian. His father had worked for a Russian concern in Qazvin (in charge of constructing the Enzeli-Qazvin-Tehran road). At age seven, the father had sent his son to Russia to receive a proper education. He had stayed there with the family of his father's associate. His name was Abdol-Samad Kam-Bakhsh.[34]

With the outbreak of the Russian revolution and civil war, Kam-Bakhsh had returned to Persia (c. 1918), his secondary studies incomplete, but was employed as translator in the Russian legation in Qazvin (1921). He then returned to the Soviet Union (c. 1922) to complete his studies. He completed his secondary education and signed up at Moscow State University to study economics and political science. But he did not finish this program and returned to Persia (1925).

This time he was employed in several Russian consular services, including its military attaché (Shiraz, Qazvin, possibly Rasht and Tehran). In 1927, he participated in a government examination to select pilots for Persia's Air Force to be trained in the Soviet Union (as several planes were bought from the Soviets). He was selected, sent there, and completed his training in exemplary manner.

In 1932, he returned to Persia. Shortly thereafter he was arrested, along with two others, on suspicion of leaking information to the Soviets. They were locked up, but freed after eight months, and expelled from the force. He found employment in an automobile factory, but this was only a cover, as he had developed a vast network over the years, not only among political activists, but also cultural circles, military personnel, and tribal figures, in several cities and regions.

We do not know how much of this background was known to Taghi Erani (the prior record of arrest as a pilot was known to him), and we do not know to what extent Taghi Erani opened up with him during the initial phase of contact (April–May 1935). Most likely, Erani treated this contact with caution.

In June 1935, Taghi Erani left for Berlin, a trip which he must have been planning for some months. He was retained by the Ministry of War to evaluate recent developments in forensic chemistry. His other justification was possibility of marriage. He passed through Baku, wrote a letter to Kamran. He indicated that he was heading to Europe and that he could pass through Moscow on his way back. He provided a contact address in Berlin.[35]

He moved on to Berlin and stayed there three months (June-August 1935), with a one-week excursion to Paris and London. We do not know what contact, if any, he was able to make there. In Germany, the left was crushed, but in France and England they were active. In France, he may have hoped to find the trails of Münzenberg and Breitscheid. But in all likelihood such contact did not materialize. In London he visited a bookshop, bought some works by Marx and had it shipped to Brukheim bookshop in Tehran. His stay in Europe coincided with important events which he reported in his journal (*Donya*), after return to Persia. In Paris, an anti-war gathering had taken place at *Palais de la Mutualité*, in which Gide, Malraux, Barbusse, Brecht, Huxley, and many other celebrities attended. They gave orations to jam-packed crowd of over three thousand (June 21). In Great Britain, a naval show of force was put on at Spithead for the silver jubilee of King George V (July 16), one hundred and sixty warships. In Berlin, a number of Jews were murdered at Kurfürstendamm (July 15); also, Jewish tourists were expelled from the seaside resort of Misdroy (July 27); he reported all these (except the seaside incident).[36]

Erani left Berlin for Moscow, arriving there during the last week of August 1935. He showed up at Gorki Avenue, Pushkin Square, in front of the statue, and met with Kamran. He brought him a winter jacket from Berlin, as per his request. He was given a package to hand over to the operative back home, the Russian-trained pilot, Kam-Bakhsh. It contained French bills, six hundred and fifty Francs. He was informed that the Comintern directive issued at the closure of the seventh congress (August 21) is to abandon all communist-related activity, including the shutdown of his journal, form associations within civil society, and persevere toward the formation of the United Front against fascism.[37]

After a three-day stay in Moscow, he returned to Persia and put out the final issue of *Donya*, as a mega issue, printing in it what he considered to be the essentials of his philosophy and outlook, not adequately covered in previous issues.[38] This issue was dated May 1935 (using the date of the approved version from the censor office, prior to the trip), but it came out in September 1935 with new content (it contained news from June–July 1935 cited above, one of his disciples, the print-setter at *Sirousse Press* slipped it through, Akbar Afshar-Ghotooli). In this issue, he highlighted the key points in boldface as concepts for the readers to retain and indicated the next series will come out at the start of the following year (in Persian calendar, March 1936), for wider readership. What led to this closure, most likely, was not only the Comintern directive, but also the fact that it had become quite taxing to sustain the publication project (the burden principally on him).

Erani then engaged in various above-ground associations in collaboration with some of the sympathizers of the journal and personal acquaintances—he joined the friends of League of Nations (he became a member, society was presided by the newly appointed Minister of Justice, Ahmad Matin-Daftari), solidarity fund for print-setters (he initiated it with a print-setter from *Sirousse Press*, as the prior fund had been misappropriated by those managing it, but this attempt did not get off the ground), lawyers for social well-being (backing a seasoned lawyer, Shahidzadeh, an orientalist buff who had assisted Taghi Erani with the Khayyam project), doctors for social well-being (led by his friend from Berlin, Mohammad Bahrami), a youth club (formed but taken over by socialites, intended to promote sports and healthy lifestyle, became a gaming den). Some of the sympathizers led a student protest against the autocratic rule of the university president, whose allegiance to Reza Shah was well known (Professor Mahmoud Hesabi, with a personal audience with the king), using pretext of new rules that revoked advantages related to government recruiting. It was in conjunction with this action that Taghi Erani told one of his former students, Anvar Khame'i, involved in the strike action—"revolutions are made by scruff-necks, *yagh'e-cherk*, not stiff-necks." Author did have correspondence with Anvar Khame'i.[39]

Taghi Erani developed a working relationship with the pilot, Kam-Bakhsh (possibly the Moscow visit gave him the false assurance that he was authentic and the right guy), with whom contact was one way. The pilot acted as the hub.[40] This was the cliff. Erani did not take his own thinking to its logical conclusion. If a party is what leads a revolution, and cadres what lead the party, is not the one mastering over the cadres the master of the situation? But now he was cut off not only from his potential cadres but also his journal and weekly philosophical gatherings; they were both shut down. He had become vulnerable, indeed dispensable. His network had been hijacked to be used as a band of informers and infiltrators, marionettes in the hands of its ringmaster, Kam-Bakhsh.

Lenin may have been deficient in philosophy of science, as Erani had hinted, or in philosophy for that matter (as Plekhanov had suggested, Erani's monism with certain similarity to that of Plekhanov), but Lenin definitely knew his politics in the most pragmatic sense—cutting deals with the Germans to reach Russia (via Baltic territory to St. Petersburg, Finland station), cutting another to prevent them from marching onto Moscow (Brest-Litovsk), cutting yet another to gain access to their chemical warfare technology (Rapallo Treaty), cutting deals with bank-robbers (Stalin and Georgian clique), and extracting money from Russian aristocrats having lost confidence in the Tsar's ability to manage the affairs (Krasin, chief operating officer), cutting deals with the British over "peoples of the East" and so forth. Other so-called greats did the same—Chinese communists cutting deals with the Russians (Mao and Stalin deals, despite Stalin's backing of the KMT that had caused a massacre of Chinese communists, then Mao cutting deals with the KMT, backed by US aid, to fend off the Japanese). Indian leaders cutting deals with the British (Gandhi's tacit support for the British during WWII in view of independence, a subdued version of his active support during WWI, and conceding at the end to the partition of India for the independence prize). Indonesians likewise went on to cut deals with the Japanese (Sukarno and Hatta team, collaborating with Japanese occupation to overcome Dutch colonial rule). Erani and his ilk were a far cry from this sort of pragmatism or Machiavellianism in political action. Even at a theoretical level, the paradigm of "vanguard party" was a rather reductionist formula for revolutionary action. But he carried on as he deemed fit and it would soon end in disaster.

Taghi Erani handed over the package from Moscow to the pilot and passed on names of various people thought to have aptitude for political activity. The two met every three to four weeks, a total of twenty to thirty times (September 1935–March 1937). Taghi Erani's residence had a telephone, and that is how the ex-pilot would get in touch with him. The *Hotel Dariush* was one of the *rendez-vous* spots. The pilot employed a smuggler to dispatch the wife and son of Dehzad to Russia. They made it across the border without any difficulty. The pilot showed to be capable of achieving results.[41]

The pilot also appointed someone he had known from Qazvin, now in Tehran, to liaise with part of the group on his behalf and operate a leased home as a motel for the other operatives. His name was Zia Alamuti, alias *Fattane*. The pilot worked with some of the members directly, including the youth. With each he had a different alias. But he led them to believe that the organization is in preparation for eventual seizure of power. They had to report about whatever they observed.[42] Taghi Erani lost his contact with his sympathizers; he only maintained a direct link with the medical doctor (Mohammad Bahrami), whose clinic was used as a hub for courier and messaging. He also kept his social contact with few others.

Then came the following sequence of events, in rapid succession—the operative back in Moscow (Kamran) wrote a memo about Taghi Erani to the Comintern, shortly after Taghi Erani transited through Moscow

(August 1935), indicating he can be a useful element (memo, September 1935). Another memo was written about him by a Comintern operative in Moscow and Persia specialist, Obreliani, with mixed reviews (December 1935). Then, in March 1936, there appeared a Comintern memo about Taghi Erani, replying to an earlier memo (not found), indicating he is no longer in the Soviet Union.[43] Prior to this came the publication in Tehran of Taghi Erani's annotation of Khayyam's critique of Euclid (February 1936, in which he paid a moving homage to Rosen).[44] In May 1936 came anonymous underground pamphlets in Tehran, translation of short Marxist texts (*Communist Manifesto*) and a May Day tract, both attributed to Taghi Erani.[45] Interestingly, the May Day tract was not in conformity with directives of the Comintern's seventh congress. It highlighted the war in Ethiopia, and how this proved the bankruptcy of the League of Nations. Italy and Ethiopia were both member countries, but one invades the other without any consequences. The British forces, regionally present, do not intervene, proof of appeasement not only with Germany but also with Italy. It also took aim at Reza Shah and British colonial presence in Persia, notably in the oil sector and railroads, built with the purpose to connect with India, a British stronghold. The tract was handwritten, and about a dozen copies were made and distributed to a small circle of sympathizers.

Then, in June 1936, Erani was appointed Vice-Minister of Industry in charge of its educational establishments (the Minister, General Amanullah Jahanbani, pushing it through with consent from the Prime Minister, Mahmoud Djam), got busy with administrative duties, site visits (schools and mines, photographs extant), and plan to replace German experts in his service by local staff.[46] *Rendez-vous* with the Russian-trained pilot slowed down considerably. However, he did meet with an old acquaintance from Berlin, in autumn 1936, on a visit to Persia to collect labor statistics (Jamalzadeh, member of pro-German party during the war, turned to journalism and literature afterwards, moved to Switzerland and worked for the International Labour Office in Geneva), in whose journal in Berlin, Taghi Erani had published two articles, and he had also been Erani's liaison officer with the embassy during the last two years of his stay in Berlin, when he received financial aid from the government. Erani passed on several issues of *Donya* to this acquaintance as a souvenir, though the journal was no longer in print (author had correspondence with Jamalzadeh).[47]

In January 1937, a political activist from Rasht (Ismail Foroohid, alias Salimi), who had been active in a local cultural association and taken in as a left-wing suspect was freed from jail. Shortly afterwards, he left the country illegally (most likely on steamer from Enzeli to Soviet Azerbaijan). But he was frightened by what he saw, notably the anti-Trotskyite hysteria and show trials, and persecution of Persia's communists that operated in the Soviet Union. As a result, he rushed back home (February 1937), but was arrested at the border. He was threatened with execution and gave a wide-ranging confession. This included information on the Russian-trained pilot (Kam-Bakhsh), who was picked up soon thereafter (April 1937).[48] As the pilot had had a prior history, he too decided to collaborate, though unlike the first, he had a deeper game plan, as he was not really a political activist, but an operative of Soviet intelligence (most likely recruited in early years of Bolshevik power, thus witness to the successive incarnations of Soviet intelligence apparatus, Cheka/OGPU/NKVD, pre-KGB).

A complicity developed between this agent and Persia's police, in essence to nail Taghi Erani, who quite recently had been invited to two high-profile court functions in the presence of the Shah (March 4 and 15, 1937, though we have no evidence that he attended).[49] The details of possible motivations for this complicity, on both sides, would emerge over time but one was protection of Soviet infiltrators in Persia's armed forces. This arrest led to the instrumentation of a smuggler, who was already in police custody (Mohammad Shureshian, after his theatrical performance in Ahvaz, play named Betrayal and Loyalty, *Khiyanat va Vafa*), to escort the police to the residence of Taghi Erani and a Berlin-educated medical doctor in Tehran (Mohammad Bahrami). This happened on May 8, 1937.[50]

The following day, the "motel" operator (Zia Alamuti, with a female alias, *Fattane*) was arrested. Then came many more arrests, including four who had collaborated with the journal (Bozorg Alavi, the aspiring novelist, Iraj Eskandari, the Grenoble-educated lawyer, Abol-Ghassem Ashtari, the primary technology contributor, and Anvar Khame'i, a former student of Taghi Erani who had produced an article on history of science).[51]

Amongst those rounded up, were also those who had merely subscribed to the journal. Indeed, of the total number rounded up (fifty-three), half were bystanders, a fourth had had only cultural and intellectual interactions, including some of the contributors to the journal, and only a fourth had attempted to act as a political organization, presumed nucleus of a future political party. In this way, Reza Shah's police outdid Stalin's (Stalin's show trials included—the trial of the sixteen, July 1936, and the trial of the twenty-one, March 1938, but in Persia, it would be the trial of the fifty-three, November 1938).

Taghi Erani was most concerned about his family. His father had passed away a few months prior to the arrest. He lived with his mother and youngest sister. The sister was deeply attached to him. She was only seventeen and had known him since the age of nine (when he left for Berlin she was only two years old). One of his testaments in the early days of captivity to several inmates was to ensure that she would be able to continue with her studies and do higher education, something which materialized. His other sisters were married and well settled. He had never engaged in political discussions or activism with his family. His first sister had made a literary contribution to his journal, but that was so benign and published without surname that it could not possibly implicate her. At any rate, he emphatically denied any family involvement in his interrogations. His mother, vulnerable, suffered from depression, according to his statements, based on visits he had from his sisters, and later from the mother (no more than seven visits, as he was at *Qasr* prison a total of seven weeks only, visitation rights were once a week, and there were none at the detention center where he was held, *Zendan'e Movaghat*, which was part of the police headquarters compound in the center of Tehran, the *Shahr'bani*).[52]

Taghi Erani assumed that the error from his doctor friend (Mohammad Bahrami), who had come to his residence with the smuggler must have precipitated this situation; although the two had spoken only in German. However, he was also Vice-Minister of Industry, so this may have arisen from a jab at his superior, to try to destabilize him by eliminating a capable subordinate (his superior, General Amanullah Jahanbani, indeed did fall within a year). It may also have originated from his own rivals to eliminate him as a contender to higher office. The situation was tense and confused. All he knew is that he ought to maintain his calm, be alert and quick on his feet to fend off any lethal arrows. He was always a fast thinker; overwork a definite contributor to some of his tactical errors. This drama would unfold over the next thirty-three months.

NOTES

1. See Chaqueri, *Taghi Erani dar Ayeneye Tarikh*, pp. 227–233, for bio of Abdol-Hossein Dehzad (1895–1942).
2. See Chaqueri, *The Left in Iran (1905–1940)*, pp. 209–213, for minutes of this gathering translated from German to English; he gives the location as Ivanova. See Cronin, *Reformers and Revolutionaires in Iran*, pp. 155–156, in article by Atabaki, he gives the location as Rostov.
3. For Erani and Dehzad connection, see Boroujerdi, *Erani Faratar az Marx*, pp. 89–90, and Momeni, *Parvandeh*, pp. 72–74, 86, 238–240.
4. Erani commenting on Dehzad, "he did not give me clear answers"; see Boroujerdi, *Erani Faratar az Marx*, pp. 89–90.
5. See Chaqueri, *Taghi Erani dar Ayeneye Tarikh*, pp. 233–241, for bio of Ladbon Esfandiari (1897–1942).
6. For Erani's synopsis of intellectual development of Persia, see *Osool'e Elm'e Rooh*, pp. 271–272, 278–281; also see the essay *Erfan va Osool'e Maadi* in Maghalat, pp. 107–123.
7. See Yushij, *Namehaye Nima Yushij*, pp. 82–87; letter written from Astara (the one in Soviet Azerbaijan), dated 1310 (c. 1931); this letter appeared in a collection of letters by Nima Yushij, renowned poet and older brother of Ladbon Esfandiari. In that work, the letter is incorrectly attributed to Nima. It should be noted, however, that Nima did know of Taghi Erani; see Ahmadi, *Khaterat'e Bozorg Alavi*, p. 143 (Nima and Bozorg Alavi were teachers at the School of Industry, formerly German School, in Tehran).
8. For Erani meeting with Morteza Yazdi, see Momeni, *Parvandeh*, pp. 50–51. Also see Erani, *Donya*, 12-issue Volume, p. 167 (footnote) in which he makes remarks about the self-centered character of those similar to the first doctor (Morteza Yazdi), without explicitly naming him.
9. For codebook, see Momeni, *Parvandeh*, p. 50.
10. For bio of Morteza Alavi, see Chaqueri, *Taghi Erani dar Ayeneye Tarikh*, pp. 241–245; on p. 243 there is mention of his female friend as undercover agent of German police. Also see Ahmadi, *Tarikhche Ferq'e Jomhooriye Enghelabi'e Iran*, pp. 31–32, 165–166. For an account of his court case, see Najmi Alavi, *Sargozasht'e Morteza Alavi*, pp. 38–44.
11. See Mirabedini, *Bozorg Alavi*.
12. For Erani and Bozorg Alavi relationship, see Ahmadi, *Khaterat'e Bozorg Alavi*, pp. 147–148; also see Ahmadi, *Tarikhche Ferq'e Jomhooriye Enghelabi'e Iran*, p. 52.
13. See Chaqueri, *Iraj Eskandari*.
14. See Chaqueri, *Solayman Mirza Eskandari*.
15. See Boroujerdi, *Erani Faratar az Marx*, p. 599.

16. See Dehbashi, *Khaterat'e Iraj Eskandari*, p. 85, about him being in the dark about the identity of CRRP principals, except Morteza Alavi.

17. See Ahmadi, *Tarikhche Ferq'e Jomhooriye Enghelabi'e Iran*, p. 17, about Eskandari's briefing at the Ministry of Justice.

18. See *Sarmayeh*, Vol. 1, preface, Tudeh, Liepzig, 1970.

19. See contrast in views of Erani and Taghizadeh. The latter promoted total emulation of Europeanism in the aftermath of WWI—"Persia must become in form and in substance, body and spirit, occidental"; recounted by Jamalzadeh; see Afshar and Dehbashi, *Khaterat'e Jamalzadeh*, p. 119. Erani, however, advocated the concept of leapfrogging; we must emulate western science and technology, but move beyond its current social model; see *Erfan va Osool'e Maadi*, pp. 144–145 (in Maghalat), also *Donya*, 12-issue Volume, pp. 9, 20, 43, 228, and in *Teorihaye Elm*, p. 321 (in Maghalat).

20. See reference to Karl Pearson (*Grammar of Science*) as an example of a circumspect and hesitant attitude with respect to science; Erani, *Teorihaye Elm*, pp. 317, 321 (in Maghalat).

21. His essay *Erfan va Osool'e Maadi* (Mysticism and Materialism) is dedicated to depicting this ebb and flow from antiquity to the present, embracing Asia and Europe.

22. See Erani, *Teorihaye Elm*, pp. 315–321 (in Maghalat); also *Materializm'e Dialektik*, pp. 30–48 (in Maghalat).

23. The conjecture that publication of *Donya* had political motives may be assessed by two accounts. One by Iraj Eskandari, how according to him Erani asked him to mentor potential recruits, recounted in Momeni, *Parvandeh*, pp. 376–379. Also in Khame'i, *Panjah Nafar va Se Nafar*, p. 89, attempt to engage with Eprim Isaac, his classmate, at the suggestion of Erani, according to Khame'i.

24. Physical description of Journal *Donya* is based on 1980 reprint of the original; color of cover being green mentioned in Khame'i, *Panjah Nafar va Se Nafar*, p. 78; journal also discussed in Khame'i interview, *Harfhaye Erani dar Iran Tazegi Dasht*.

25. Distribution in Abadan is inferred from remarks by Khame'i that Mostafa Fateh (in APOC senior management), was a supporter of the journal, and did not attempt to block its distribution; see Khame'i, *Panjah Nafar va Se Nafar*, p. 19. Distribution in Shiraz is inferred from technology contributions of Abol-Ghassem Ashtari who was vice-principal of a vocational-technical school in Shiraz. Distribution in Tehran is well established by commentaries from Bozorg Alavi, Iraj Eskandari, Anvar Khame'i. I obtained the circulation figure estimate of two hundred from Bozorg Alavi (private communication). Also inferred from money owed by Taghi Erani to *Sirousse Press* (40 Tumans, each issue costing 2 Rials);

see Momeni, *Parvandeh*, p. 239. By comparison, circulation figure of national daily (*Ettelaat*) was five thousand.

26. Anvar Khame'i states Hossein Afshar was a Majles deputy (in *Panjah Nafar va Se Nafar*); Bozorg Alavi contradicts this in his memoires, see Ahmadi, *Khaterat'e Bozorg Alavi*.

27. See Erani, *Shesh Mah Majel'e Donya va Enekas'e An*.

28. See Boroujerdi, *Erani Faratar az Marx*, p. 95.

29. See Momeni, *Parvandeh*, pp. 74, 240.

30. See Ahmadi, *Tarikhche Ferq'e Jomhooriye Enghelabi'e Iran*, pp. 70–72; also described in Jamalzadeh correspondence in Golbon and Sharifi, *Mohakemeye Mohakemegaran*, pp. 201–203.

31. See Momeni, *Parvandeh*, p. 241, on Mrs. Kamali and Mr. Shirinov; and p. 52, on Reza Radmanesh. Latter is also described in Khame'i, *Panjah Nafar va Se Nafar*, p. 110, as well as in Ahmadi, *Tarikhche Ferq'e Jomhooriye Enghelabi'e Iran*, p. 153, based on memoires of Avanesian.

32. See Momeni, *Parvandeh*, p. 240. KUTV is the designation for this university.

33. See Momeni, *Parvandeh*, pp. 76, 78, 95–100, also Ahmadi, *Tarikhche Ferq'e Jomhooriye Enghelabi'e Iran*, pp. 72–73. For bio of Nosratollah Aslani see Chaqueri, *Taghi Erani dar Ayeneye Tarikh*, pp. 249–253.

34. For bio of Kam-Bakhsh, see Chaqueri, *Taghi Erani dar Ayeneye Tarikh*, pp. 210–226. See also Khame'i, *Panjah Nafar va Se Nafar*, pp. 30–62.

35. See Momeni, *Parvandeh*, pp. 242, 248, 252.

36. See Momeni, *Parvandeh*, p. 229 (Paris), p. 246 (London), p. 247 (Moscow). For events in Europe during that summer, see Erani, *Donya*, 12-issue Volume, p. 386.

37. See Momeni, *Parvandeh*, pp. 242, 252–253.

38. Ibid., pp. 86, 90, 242.

39. Ibid., pp. 51, 66, 86, 91–99, 239; also Ahmadi, *Tarikhche Ferq'e Jomhooriye Enghelabi'e Iran*, p. 62; also Khame'i, *Panjah Nafar va Se Nafar*, p. 81.

40. See Momeni, *Parvandeh*, pp. 79–82; Erani describing Kam-Bakhsh: "For me, he was the hub" (p. 82).

41. See Momeni, *Parvandeh*, pp. 243–244.

42. See Khame'i, *Panjah Nafar va Se Nafar*, pp. 30–62.

43. For Comintern documents on Erani, see Chaqueri, *Taghi Erani dar Ayeneye Tarikh*, pp. 87–126, 158–179, 254–259, and 263.

44. See Erani, *Resalat fi Sharh ma Ashkal men Mosaderat Kitab Eqlidos*.

45. See Momeni, *Parvandeh*, pp. 299–301. In his memoires, Bozorg Alavi indicates these pamphlets originated from Taghi Erani; see Ahmadi, *Khaterat'e Bozorg Alavi*.

46. See Momeni, *Parvandeh*, p. 245; also Ahmadi, *Khaterat'e Bozorg Alavi*, p. 147.

47. S. M. A. Jamalzadeh, personal communication.

48. Ismail Foroohid's pseudonym was Salimi; they came to be called the group of "twenty-three"; see Boroujerdi, *Erani Faratar az Marx*, pp. 216–220, and Fami-Tafreshi, *Polis'e Khafiy'e Iran*, pp. 246–247.

49. *Erani Family Archives.*

50. Momeni, *Parvandeh*, pp. 225–228.

51. Ibid., pp. 47–56.

52. Ibid., p. 245; also Boroujerdi, *Erani Faratar az Marx*, p. 190; Alavi, *Panjah'o Se Nafar*, p. 73.

CHAPTER 14

The Geopolitical Game

When the Big Three (USA, UK, Soviet Union) met in Tehran to discuss the opening of the Western Front against Hitler, something which materialized within six months (D-Day, June 1944), Roosevelt asked Churchill: "so what finally became of Reza Shah," to which Churchill gave the curt reply—"*we brought him, and we took him!*"[1]

Reza Shah had had a notorious intelligence service, but oddly his intelligence was strictly pointed towards the domestic domain. He had virtually zero intelligence on what was happening abroad, even near his borders. A few months before the king fell (September 1941), a high-ranking official of the TIRR (trans-Iranian railroad) indicated that he had noticed large contingent of Indian infantry and British armaments at the port in Basra, a major military buildup, although Iraq was already in British hands (collapse of Rashid-Ali *coup*). He reported this to the king's secretary in Tehran and dropped by a second time to ensure this was passed on to the king, of which he was reassured. But even if the information was passed on, nothing was done about it, and indeed nothing could be done about it. It was too late.[2]

When Hitler pounced on Russia (June 1941), the fate of Reza Shah was sealed. Not only had he flirted with Hitler (even after appeasement was over, Churchill premiership, May 1940), and bargained for more money with the British over oil royalties after the war had begun (September 1939 start of war, and his haggling, in 1940), and that for

© The Author(s) 2019
Y. Jalali, *Taghi Erani, a Polymath in Interwar Berlin*,
https://doi.org/10.1007/978-3-319-97837-6_14

purely personal gain, he was also totally spent, physically and politically. He was in his seventies, bad teeth, ulcers, opium addiction, and detested by his people and the country, which he had treated as his personal possession, extorting money and property even from its elite and terrorizing the population, eliminating in the process a good part of its luminaries.[3]

Further, Persia was now badly needed as a corridor (starting from its ports in the Persian Gulf, linked to the Caspian Sea by the railway, with ferry and land access to Soviet territory) in order to dispatch armaments and provisions to the Red Army to fend off the Germans, for if Russia fell, Germany would be unbeatable: the specter of a German Europe, Europe a German Commonwealth.

This was the Lloyd George doctrine minted during the previous war that Germany plus Russia is unbeatable (knowhow in one, natural resources in the other), and it is what gave ammunition to Churchill to do away with the appeasers ("guilty men"), for once the war had begun, and the weaklings had fallen (Belgium, France), Hitler would target either England or Russia, both equally unacceptable. Thus, he had to be taken on, May 28, 1940, the pivotal date, when Churchill prevailed over Halifax in the war cabinet with the backing of Chamberlain. And this same doctrine is what made total possession of Persia a matter of existential necessity, for how else to defend Russia (Archangel, mostly frozen, Vladivostok, a logistical nightmare). And it so happened that the removal of Persia's king could go a long way to mitigate the humiliation of invasion and its illegality, for Persia was a neutral country in the conflict.

True, Persia was full of Germans, in Allied eyes (though no more than a few thousand), and possibly many spies among them (Abwehr agents), but certainly not nearly as many as the Brits and the Russians had in Persia, and for much longer. Nevertheless, the removal of a tyrant could go a long way as a symbolic act. Those who had brought him in were good enough to take him away.

In a space of three weeks, dozens of towns and cities fell to the Russians in the north and to the British and Commonwealth forces in the south, American troops joining after a few months. Domestic fatalities, at least officially, stayed under one thousand. (The author's father was on the side of defense, facing the Red Army, lasting three days before running for his life, while general Kozlov and their one thousand tanks rolled into Persia, the terrain razed by bomb-dropping airplanes; start of famine; bread a scarcity and even then, adulterated with sawdust.)[4]

While the country stayed occupied by the Allies and Reza Shah adapted to his forced retirement in exile (his family had wanted to go to Argentina, for it was vast and great for hunting, reminiscent of north Persia, but the British preferred their own turf; certainly, he should not fall into German hands and made out into a devout Muslim sovereign and victim of British imperialism, so he ended up in Johannesburg after short stints in Mauritius and Durban). While all this was going on, the local clans at home had a go at one another to settle old accounts.

The newly appointed Minister of Interior was the one Reza Shah had sacked and thrown in jail (he had been Minister of Industry, General Amanullah Jahanbani, Taghi Erani's superior, and was imprisoned shortly after Erani's arrest). Even the new Minister of Justice (Majid Aahi) had done time in a Reza Shah prison. As well, came forth powerful plaintiffs from the Teymurtash clan, the fallen Court Minister, the Bakhtiyari clan (next of kin to the murdered Sardar Asad), and others, including the family of the renowned poet Farokhi-Yazdi, and the Erani family (represented by the mother).

This caused a senior judge (Jalal Abde) to mount a case against Reza Shah's band of executioners. These included his Chief of Police, whose main job was intelligence construction and elimination of real and presumed adversaries. Also included was the head of the notorious *Qasr* prison, in the outskirts of Tehran, and the chief executioner shuttling between *Qasr* and the detention center (*Zendan'e Movaghat*) in the capital, far worse than *Qasr*, the very incarnation of doctor death. He was skillful in concocting and administering lethal potions, air injection, asphyxiation, and the like.[5]

Many others were rounded up as well: those who had played a role in high-profile executions in the provinces (notably in Khorasan, eliminating Modarres, the flamboyant cleric and *Majles* deputy, and Mohammad-Vali Asadi, Grand Administrator of the Mausoleum of Imam Reza), as well as interrogators and petty accomplices, and the predecessor of the head of *Qasr* prison, who was not as notorious as his successor, but still a despicable fellow. However, there is no mention of the arrest of the head of the detention center.

The predecessor of the Chief of Police had been the architect of Reza Shah's intelligence apparatus. He was a former Cossack like the king himself, and a relative of the king (he was direct cousin of the king's fourth wife, mother of the crown prince). During these arrests, he was beyond reach as he had fled Persia while Reza Shah was at the height of his powers (1934). By faking the collapse of his vocal cords, he obtained permission to leave the country for medical treatment and headed to Germany, but never returned. He attempted to interest the Germans in wholesale takeover of Persia and collaborated with the Third Reich to bring this about, but the Allied invasion made it an impossibility, so he exited to Lichtenstein, before the fall of Hitler.[6]

A full account of these trials has yet to be published, but two veterans of Persia's tradition of independent historiography put out a partial account shortly after the 1979 revolution, the end of the Pahlavi dynasty, a dynasty of two kings. As they put it, they felt compelled to offer what they could (including an interview with the magistrate who had led the campaign to round up the executioners), although it is well known that in Persia, those who pick up the pen do so by putting down their lives, for who knows who will be on the top tomorrow, maybe the one who has fallen today (stroke of pen, stroke on one's life—*ghalam-zadan, jaan-zadan ast*).[7]

At the conclusion of the trial, the Chief of Police got a ten-year sentence, the chief of *Qasr* prison got life in prison (his predecessor got six years), and doctor death got the firing squad. However, while the last was carried out, the others were not, at least not in earnest, for they were pardoned by royal decree within a few years of sentencing. The proceedings lasted from July 1942 to May 1943, as there were several filings. Initial sentencing fell on September 1942, a year after the fall of Reza Shah. The verdict pertaining to Taghi Erani came in May 1943, upholding the earlier sentences.

In fact, all along the young king and his court entourage attempted to prevent the prosecution of the executioners. The new king was initially weak, and only later when he gained more clout was he able to undo these sentences (in 1947, after his father passed away in exile, July 1944). As a result, they were set free, and one of them, the Chief of Police, became the head of the Pahlavi family real-estate trust (assigned nominally to the king's mother), which pooled the estates that had been exacted from various proprietors during the reign of his father.

Compared to his previous job, masterminding the elimination of key figures which involved a great deal of fabrication and intellectual exertion, this new job was a breeze, giving him plenty of time to indulge in his personal interests, poker being a favorite pastime and music a passion.[8]

Undeniably, he was a skillful composer and violinist, fond of melancholic tunes, overheard not only by his neighbors, a psychopath soothing his soul by caressing the fiddle, but audible also to the whole nation. His music was aired on the national radio without disguising the identity of the artist. Such was the final outcome of a vast operation, which had had him picked up at Kermanshah (his birthplace) before he crossed the border into Iraq. Doctor death had made it across the border but was spotted and repatriated to stand trial. In similar vein, the Minister of Justice under Reza Shah was incarcerated by the Allies on charges of being a Third Reich fifth columnist. It was he who had orchestrated the Erani show trial in complicity with the Chief of Police. He had menaced the defendants' lawyers who had shown a semblance of courage (two of nine) and slandered the families of the accused, for having brought up such rotten offspring. For this performance, he was rewarded by becoming Prime Minister shortly after the trial. The Allies locked him up for a short while, but at the end he was let go. Such was the aftermath of this affair. Now, we may probe the causal chain which led to the Erani Affair.[9]

THE CAUSAL CHAIN

Reza Shah was never too impressed by the Weimar Republic in its handling of relations with Persia. They were responsive and would do their best but could not guarantee results. This caused a breakdown of relations between the two countries in 1932, when the Weimar courts ruled against the government of Persia in the case of a left-wing journalist (Morteza Alavi, one of the CRRP founders) who was spreading poison about his rule, so they claimed. And this was at a time when the German share in Persia's trade had approached double digits in percentage terms, still a far cry from the position of Russia and England, but it had begun to close the gap.

This situation changed with the Third Reich, as even German nationals could not contemplate left-wing politics, much less foreigners on German soil. And those who were suspected of left-wing leanings would be immediately expelled. This happened to yet another student from Persia in May 1933, and a few others who had been KPD members during the Weimar era had to tear up their cards and keep a low profile. In France and England matters stood differently, the left was free to function, but that is how it was in Germany.[10]

It must also be said that Reza Shah came to power rather late, he was fifty eight at his coronation (though forty eight officially) and had begun life in abject poverty. He had an accentuated urge not only to indulge in self-enrichment, but also to project his dynasty far into the future. (He was his father's thirteenth child, his mother's third or fourth, and the only child of their joint wedlock. He lost his father at birth and was left behind by his mother while he was in his early teens. He was cared for in a makeshift orphanage. Then, before reaching legal age, he enlisted with the Cossacks, taken in as a probationer. These are based on memoires of old Cossack officers, who recount his further hardships.[11])

Reza Shah's two objectives were somewhat at odds, self-enrichment and founding a long-lasting dynasty. Laying the foundation of a dynasty that could rival the Achaemenid or Sassanid (for his reference was ancient Persia) required an economic system, infrastructure, and institutions that would outlast him and his offspring. These would have to be financed by whatever means the country had at its disposal—oil rent and proceeds from a variety of concessions, some due for renegotiation, trade tax (tariffs), and wealth tax (principally on property, hence land distribution a necessity, as it was monopolized by a minority). Also, revenue tax would have to be used based on a progressive taxation regime to maximize yield, and aggregate savings stimulated (regulation of wages and prices). Government proceeds could then be deployed for investment projects, complemented by borrowing from the capital markets, to develop the home market and expand exports, as opposed to capital-intensive projects for foreign-influenced military prerogatives, railroad with geostrategic itinerary, and deepwater ports for a country with no merchant shipping or warships. Given his grip on power, virtually a monopoly, he was in a position to achieve a good deal of such economic development and growth agenda.

However, these ran counter to financial gain in the short term, not only for the royal family, but also for the entire landowning establishment and moneyed classes, which had vested interests in preservation of serfdom, avoidance of wealth and revenue tax, and exclusive recourse to taxation of consumption. For the innermost circle, there were also the proceeds of oil revenue diverted to foreign accounts, skimming aggregate savings by the devaluation of currency, taxing merchant capital by engineering the fluctuation of exchange rates, and indulgence on the account of the state (palaces, court extravagance). In short, what was in vogue were short-term methods aiming to achieve a long-term goal, an untenable program, yet with a fixation on longevity.

This obsession with survival, gaining time and endurance, meant that nothing could be taken for granted. It was better to strike when in doubt than wait for proof and discover that it was too late to save the situation. In this way, Reza Shah managed to keep one step ahead of his adversaries, real and presumed. This principle of preemptiveness was not limited to domestic affairs, it also operated at the level of foreign relations and geopolitics.

On that front, Reza Shah's greatest fear was Russia. He did not trust the British either, despite the fact that they had done more than anyone to help his accession to the throne. But as far as he was concerned, they had been amply rewarded for this investment. They kept their naval base in Bushehr and gained access to a second port in the Persian Gulf, brand new, paid for by domestic funds, and much better suited to warships. Port Shahpur was a deepwater port with direct access to the trans-Iranian railroad, TIRR. The TIRR was also built during his reign, with domestic financing, with ready access to Russia, in case of war. Its terminus was Port Shah on the Caspian Sea, near Shah's estates in Mazandaran province, and would be connectable to the Indian railways as a future extension possibility. It was also a ferry ride away from Soviet territory, Turkmenistan.

The British had also kept their monopoly over Persia's considerable oil production, paying Reza Shah a modest profit-sharing fee (sixteen percent of APOC after-tax profits, that is profits after taxes were paid to the British government, a payment which far exceeded Persia's royalties), about one million pounds sterling a year. Post-1933, this doubled as they moved to the royalty of four shillings per ton, about five pence per barrel. This sum Reza Shah kept abroad. It was a personal account as it never entered Persia's national budget, to the consternation of Mr. Millspaugh (an American, Persia's financial controller, 1922–1927), though occasionally military hardware was bought by it.[12]

Thus, with the British he kept a business-like relationship and never discounted the possibility that one day they would drop him for someone else. It was in a way a saving grace that the successor turned out to be his son. But that was not done to ingratiate Reza Shah. It simply turned out to be the least bad option for the British at that juncture. It could very well have been someone else. As Sir Reader Bullard told Reza Shah—he had no say in the matter, he simply had to go (similar to what General Huyser told the last Shah, a terse expression of US sentiment, before he exited the country in 1979). Thus, relationship with the British was business-like, give-and-take, at best one of cautious optimism.[13]

Russia though was a different story. First of all, their methods resembled Reza Shah's the most, and not by chance. He had slaved under them for over three decades, initially as a probationer and eventually as an officer of the Cossack Brigade. He reached that rank at the *Otryad* in Hamedan (though always subordinate to a Russian officer, who were under command of a Russian general, accountable to the Russian Minister in Tehran and to St. Petersburg, though nominally they were accountable to the Qajar monarch). This brigade was orphaned after the Russian revolution, and that is how General Ironside was able to decapitate it without a single shot being fired. (General Starosselsky's telegram to his men while he was on his way to meet with Ahmad Shah in Tehran was intercepted, and doctored, ordering his men to arrive at a camp near Qazvin, which was under control of Norperforce. Upon arrival, the Russians were disarmed, dispatched by lorry to Basra, and consigned to a ship heading to Vladivostok, delivering them to the white army, who were fighting off the Bolsheviks.)

Following this decapitation, Reza Khan was put at the helm and instructed to march on Tehran, as the surest way to ensure British influence on Persia after the departure of Norperforce (while British forces in the south never left, Bushehr was always in British custody, and Khuzestan, the oil province, virtually a British mandate). Five years elapsed between this event and his coronation (1921–1926), during which he showed to be a superb tactician on both military and political fronts—ruse, dissimulation, waiting games, reliance on turncoats, coercion, intimidation, and so on. This success brought about increasing support from the British and caused a small group of Qajar-era statesmen, ambitious and somewhat naïve (for nearly all of them were devoured by their master), the Young Turks of Persia, to gravitate towards him, forging an alliance of force and finesse, epitomized by the pact between Reza Shah and Teymurtash.

Toward Russia, the duo adopted an approach marked by cooperation and precaution. This was the only sensible course of action as Russia was Persia's largest trading partner and was destined to remain so by virtue of its vast transport routes (land and water, latter via Volga-Caspian waterway). The northern customs was a significant source of the state's revenue in a country that lacked a system of taxation. But they had to be cautious as well and prevent a Soviet spillover into Persia, as otherwise the British would be the first to remove them. They were there to act as a dyke and as long as they fulfilled this function effectively, without encroaching on the strategic interests of the British Empire, they could enjoy its windfalls.

The Russians, on the other hand, saw northern Persia as an extension of their territory and made sure to have the logistical means to appropriate it as need be. This included not only preparedness on the Soviet side, but also receptiveness on the Persian side—allies and agents, recruits in the military, tribes, administration, media, and infiltrators among intelligentsia and worker's associations (in the guise of indigenous political and trade-union activists).

This caused a number of incidents and clashes between Russia and Persia, when Persia would attempt crackdown on these activities. This included a sudden revoking of transit rights for Persian goods, blocked delivery of kerosene into Persia (ironic for a country rich in petroleum resources), arbitrary augmentation of tariffs, expulsion of Persia's migrant workers in the hundreds, amongst whom native agents could be mixed in, border skirmishes, unrest in northern provinces and even deeper inside the country, well into British turf. When they wanted to, the Russians could play hardball and make life hell for the Persians. While the clique in power would try to patch things up, but this was not always possible, as they also had to fulfill their obligations towards the British.[14]

When Bajanov defected by entering Persia (1928), Stalin intervened personally, offering to turn over control of the Caspian fisheries concession to the Persians. But his request could not be satisfied (and his offer was certainly not genuine), and Bajanov was turned over to the British in India, from where he went to London for debriefing. Likewise, in the wake of Agabekov defection (1930, via Turkey to England then Belgium), some two hundred suspects were arrested in Persia, and one was immediately shot. This was a blow to the Russians, as some of those rounded up were certainly their agents. Overall, it was an uneasy relationship. It is no wonder that when the Allies invaded Persia, a blitz operation over in three weeks (though they stayed for five years), Reza Shah was most apprehensive about the Russians, constantly on the phone to see how far they had advanced for he was certain that they held a special grudge against him.[15]

When the Russians started approaching the capital (the understanding with the British was that they would go no further than Qazvin, but anything was possible, a commando raid), he left without delay, heading south, into British-controlled territory, along with his entourage (eleven family members and as many staff). He reached Bushehr, sailing from there to Bombay, transshipping to a larger vessel and on to Mauritius in the southern hemisphere, where soon it would be summer, with a climate worse than south Persia, hence the commotion to get out of there. This led to his relocation to Durban and finally Johannesburg (in a rental property belonging to a Jewish doctor, the neighbors cross with his boys, for they were unruly, the father unable to rein them in. Surely, they must have wondered how he had managed to rein in a whole nation for so long).[16]

***

Before this exit, however, during the last seven or eight years of his rule, an important development took shape, which made life more tolerable for Reza Shah and his entourage, giving them some respite from the tit for tat in which they were caught, Russians in the north, British in the south, one feisty the other exigent. This development was called the Third Reich. And it is this which lies at the heart of the Erani Affair, in an insidious way.[17]

German trade with Persia skyrocketed during the Third Reich. It had reached about ten percent at the end of the Weimar era, then came a halt. It restarted in the early Hitler years and reached nearly fifty percent by the time the Allied invasion took place. All this is well known. What is less known, or less commented upon, is that this soaring trade happened with full cooperation of Russia and England. Indeed, there were no other trade routes to Persia (barring Turkey, and the Black Sea route, of which German sea-liners did take advantage, but it was minuscule, as was Turkish trade with Persia, and this time around the Turks were circumspect when it came to the Germans, and when conflict broke out they sided with the Allies).

The Russians gave access to German trade with Persia (via Volga-Caspian route, as well via land routes), as their version of appeasement, which culminated in the Molotov–Ribbentrop pact, August 1939. The expectation was that this would encroach on the British market share in Persia. The British, likewise, gave access via the Persian Gulf (Hamburg–Khorramshahr an important route), calculating it could encroach on the Russian market in Persia, and it was aligned with the overarching strategy of appeasement (enshrined in the Munich Agreement, September 1938, which operated until Churchill took the helm in May 1940). Of course, the Russians and the British both collected their transit fees, but were it not for geostrategic considerations, such fees could never justify losing a monopoly position to a competitor known for its resourcefulness.

The ruling elite in Persia, as well, saw this as a green light to keep going with the Germans. Various companies homesteaded in Tehran and the provinces. Trade missions arrived, and credit facilities were established in the form of a barter-based clearing agreement, as the Shah had still not given up on his vision of founding a long-lasting dynasty. Even if that did not materialize, the increasing trade volume was that much more opportunity for commissions and kickbacks. He was after all the principal gatekeeper.

The added bonus was that this relationship fulfilled the principle of precaution. This was central to Reza Shah's mantra of statecraft and preservation of political power. Here was another force, one far more willing than the British seemed to be, to take on Russia, which in his mind was *the* existential threat to Persia. Russia was a threat to his rule and his vision of a dynasty that would catapult him to the ranks of a Cyrus, Dariush, Ardeshir, Khosrow, and other immortals of Persia's royal pantheon. He was offended when a panegyrist likened him to Nader Shah, certainly one of Persia's most accomplished kings, saving her from Afghan and Ottoman onslaught, as well as elbowing the Portuguese out of the Persian Gulf. But that was not good enough for Reza Shah, and he dismissed Nader Shah as a bandit.[18] As far as the precaution prerogative was concerned, Reza Shah imagined he could do no better than to have both the British and the Germans as backers against a Russian onslaught which was sure to come. Little did he suspect that the Lloyd George doctrine would get a second life, making not only Russia but a Russia backed by England, the existential threat to all that he had fathomed, which turned out to be nothing but pipedreams.

The relationship between the Persians and the Germans climaxed with the visit of Schacht to Persia. Schacht was the mastermind of German economic planning during the formative (pre-assault) phase of the Third Reich. (He had been a Weimar technocrat before that.) He arrived in Tehran in November 1936, where he shared his vision of a collective autarky: a division of labor between several countries synchronizing their economies. On the ground in Tehran, during his visit, there was a large contingent of German notables. These were not only government and diplomatic officials, but also industrial lobbyists with an inside track that was much more efficient than channels available to official representatives of the Reich, notably the Leese-Davar linkage (as elaborated by Professor Jenkins), former representing a powerful consortium of German industrialists, latter the Finance Minister. Schacht was a big hit in Tehran, and while the Shah played his usual tactics, leaving for the Caspian as soon as the Third Reich economic Tsar landed in Persia, he nevertheless journeyed to Rasht for a personal audience with the king, and from then everything went into overdrive.[19]

Then came the shake-up at the Ministry of Industry, which seemed to be going against the tide, the fall of Taghi Erani in May 1937 (Vice-Minister since June 1936, drafting his substitution plan to get rid of German experts in his jurisdiction, in a space of two years) and the fall of his boss, the Minister of Industry (General Amanullah Jahanbani), whose military credentials did nothing to spare him.[20]

But there was more to this story, indeed several layers more, as at precisely this time the Erani political circle had begun to spin out of control. A left-wing activist known to the police in Rasht (Ismail Foroohid, alias Salimi), disappeared from site, shortly after he was released from prison. He had crossed the border into Soviet territory. Having been out of touch with events in Russia, he found to his great horror a witch hunt not only at the pinnacle of power (hunting "enemies of the state" and Trotskyites), but also at the base of the pyramid, virtually the whole leadership of Persia's communists were locked up, exiled or eliminated. Those who survived were either a part of the Soviet intelligence apparatus or possessed some particular niche or identity (such as the poet Lahuti, a cultural celebrity in Tajikistan), or were simply forgotten in some Soviet Gulag, and would surface years later. The rest were normal civilians, apolitical, migrants from Persia or local-born descendants of migrants. As a result, he rushed back to Persia (February 1937), where he was picked up by those who had set him free not long ago. This led to a wide-ranging confession, in which a certain Russian-trained pilot was named. As the police already knew about the pilot, they spotted him, and he was picked up within weeks (Abdol-Samad Kam-Bakhsh, April 1937). Shortly before this, Taghi Erani was invited to the *Golistan* palace for the king's birthday ceremony and the joint wedding of the king's two daughters (both events in March 1937, invitations issued by Mahmoud Djam, Prime Minister, whose son was marrying one of the king's daughters). Taghi Erani had the right profile to occupy high office at a time the German–Persian relations were reaching its climax. But at this time a smuggler the pilot had been using was already in police custody (Mohammad Shureshian, February 1937), and he knew not only Taghi Erani by sight, but also his place of residence, having accompanied his handler, a Berlin-educated medical doctor (Mohammad Bahrami), to the Erani residence. The smuggler was in police custody as he had committed a gaff in his last mission, bringing two operatives into Persia (Boghrati and Sadeghpour), not immediately switching from their camouflage gear, Red Army uniforms, into local clothing. Therefore, they were spotted on the Persian side of the border, arrested, but fled, though left behind their papers. This led to the smuggler's arrest in Ahvaz where his brother had given him shelter while he dabbled in theater for working-class folks, using his own name. And the confluence of these—the critical element being the extensive cooperation of the pilot with the police, the motivation for which remained fuzzy—led to the arrest of Taghi Erani in Tehran (May 1937) and many others, fifty three the final tally.

This development was rather fortuitous for the government, for the last thing they wanted was to announce discovery of a group, whose ringleader was a Russian-trained pilot, though native of Persia, with a history of attempted espionage, hence, certainly an NKVD agent, leading to souring of diplomatic relations with Russia, hardball retaliation, and unnecessary hardship for the government of Persia. One option was to lock up the suspects and not make any waves, no declaration, no court trial, as was the customary practice. But in this case a better option appeared on the horizon, to those at the helm, which caused a break with tradition.

It was deemed better to publicize the event, but zoom on someone else as the leader, no highlighting of foreign intervention or espionage (certainly none in the final sentencing), ideally not even mention of the NKVD agent, rather prosecution of natives, not connected to any foreign power, on grounds of violation of the law of the land. This was the 1931 bill banning dissemination of socialist and communist ideas; a legislation which was illegal in view of Persia's constitution (which guaranteed freedom of expression and association, elaborated in the original text of 1906 and its addendum of 1907). However, that did not matter, as everyone knew that the constitution was there only for lip service, as was the *Majles*, a forum for rubber-stamping the king's dictates. And there was indeed a character that fit the bill perfectly, well suited for what the government contemplated, identified in a chain of intelligence operations, someone by the name of Taghi Erani.

A number of factors made Erani the person most fitting for the role envisaged by the authorities. He was against those in power and had clearly wanted to bring about the downfall of the regime, as articulated in the CRRP literature. This required severe punishment.[21] He was also an insider, making him the biggest catch in the repertoire of the Police Chief (Mokhtari), who was no match in this respect for his predecessor (M-H Ayrom), the one who had brought down Teymurtash. Erani had also ruffled feathers of certain cabinet ministers, particularly the Minister of Justice (Ahmad Matin-Daftari, who had been present in meetings with the Germans, as he was German speaking, by virtue of his secondary schooling in Tehran, higher education done in France, the two having crossed paths in these events and in the society for friends of the League of Nations). Erani had also disenchanted the Minister of Culture (Ali-Asghar Hekmat). The Minister was proud of his erudition concerning Persia's ancient history, the magnificence of its past and the genius of its poets, and had written copiously on the matter, possibly one of the reasons he was appointed minister. Erani had dispelled these as historical nonsense, in his hard-hitting journal (*Donya*). Some influential figures had tried to put an end to Erani's journal, but Erani was rather resourceful with a good following, including amongst certain members of the cultural, political, and administrative elite. It had proven no easy matter to muzzle him.

But now with the German bonanza came a golden opportunity. A show trial of natives branded as communist would demonstrate that the affinity between Persia's elite and the Third Reich was not limited to economic and commercial considerations, but went much deeper into race, heritage, and ideology, a bond of brotherhood inked in blood, Aryanism in action. The principle of extreme precaution, covering all the bases, seems to have reached its apotheosis, *Realpolitik* at its finest.

Thus, to cement this alliance, the Reich-Pahlavi bond, what better way than to showcase the trial of a group led by an intellectual who was not only a confirmed communist (a creed outlawed in Persia, as in Germany), and which could be proven by virtue of the books found in his possession (Marx, Engels, Bukharin and others, why else would he indulge in such literature), but he was an enemy of Aryanism, indeed an enemy of racialism, taking aim at no less of a figure than Alfred Rosenberg, the guru of racialist theory, highly regarded in the Third Reich.[22]

Further, Erani was well known amongst the German expatriate community in Tehran (ensuring news of his incarceration would rapidly spread to Germany), and he was also known to certain notables back in Berlin, where he had spent several years during Weimar, engaging in science and intellectual pursuits. He had even made visits to Berlin during the Third Reich, in one as a technical representative of Persia's Ministry of War. These ensured that his prosecution would not pass unnoticed in Germany.

This mode of handling the security and counterintelligence situation that had surfaced had other advantages as well. It would send a clear signal to the nation that no matter who you are, what you do, and where you live, we are watching you (for there was enormous diversity amongst those rounded up). This would be a stern warning which necessitated that the number of accused be swollen, pulling in bystanders, guilty on grounds of proximity with the culprits, including subscribing to their journal, or contribution to it, even if related only to science and technology.

The anti-communist slant on clamping down on working-class activism would not be displeasing to the British either, who were paranoid of left-wing infiltration among the APOC workforce. The specter of the 1929 strike was still alive in the spirits, a walkout of nine thousand workers, a glimpse of what came to pass after the fall of Reza Shah, the demise of APOC, the lifeline of the British Navy.

In December 1937 came the official arrest warrant of the "fifty-three" and legal charges brought against them. This was seven months after the actual arrests, and only two weeks after Baldur von Schirach passed through Tehran. He was the head of the Nazi party youth organization, who was impressed by the youth leadership camp that was set up in Persia: the scouting organization, which functioned along similar lines to the program in Germany. There could be no better sense of timing to ensure that this news was broadcast to its intended audience. (Baldur von Schirach had a direct link to the Führer in the Nazi organization. He was also married to the daughter of Hitler's personal photographer.) This was news with the subliminal message—Persia and Germany are far more than commercial partners.[23]

And indeed, not Persia, but Iran, the land of Aryans, which at Reza Shah's initiative (triggered by the suggestion of a German minister to Persia's ambassador in Berlin), it had become the official designation of the country. And in its capital, Tehran, its central train station, designed by a German architectural firm (Philipp Holtzmann), was given an impressive ceiling carved with interlacing swastikas (the author has vivid memories of this for it was intact even in the 1960s). This implicit message of solidarity between the Pahlavi and the Reich would be reinforced at the trial, a first in Persia, for there had never been trials for those charged with political offense, certainly none charged with adhering to a particular political ideology. It would be broadcast that this is a band of card-carrying communists, a treacherous creed forbidden in Persia (as it is in Germany). The whole thing had the ring of the inquisition to it. At any rate, it would be an event not to be missed, *the* thing in the offing in the Pahlavi court, many ministries and administrators working around the clock to bring it to fruition; a promise solemnly fulfilled (November 1938).

THE SHOW-TRIAL

There was much Qajar legacy during the reign of Pahlavi. This manifested itself in the venue chosen for the trial of the "fifty-three," a Qajar edifice with ornate artwork adorning its arched ceiling, a depiction of battlefield drama in bold strokes of red and turquoise, which stood in sharp contrast to the key aspect of the whole affair that the government wanted to sweep under the carpet—the Russian element. That this was no trial to point fingers at Russia's meddling in Persia's affairs, but simply a case of homegrown bad apples brought to justice. The leitmotif of the artwork, however, was the total opposite—Russian malice and atrocity towards the Persians, the Caucasian conflicts of early 1800s. Those who had set the stage had not completely thought it through, or so it seemed.[24]

The venue had been the *haram* of the second Qajar monarch, Fat'h-Ali Shah (r. 1797–1834), known for indulgence (he was also an able poet, *Khaghan* or *Lord* his penname). This was in stark contrast to the founder of the dynasty who had lived a life of frugality, his court the battlefield, known for his extreme cruelty (he is the one who restored control of Caucasus to Persia and made Tehran the capital). But the founder could bear no offspring, as he had been castrated in childhood by a rival tribe, a practice not uncommon amongst feuding clans, and the tradition of temporary kidnapping. As a result, he picked his nephew to succeed him, which is what happened when he was murdered in the Caucasus.

The nephew was no stranger to battlefields, but for him that would not be sufficient to ensure an enduring reign. Thus, he strived to bear as many offspring as possible, which he would implant in the provinces, grooming them for leadership, while consolidating his rule. In this way, he came to have a *haram* of some seven hundred concubines from all over his realm. The great majority of these were not of Qajar lineage; hence, their offspring were precluded from contending the throne, though they could hold senior positions. This still left a good number which could fight it off for succession.

This monarch, however, happened to have a son of Qajar mother-hood. He was quite capable and his eldest, so he was anointed as the crown prince, to the chagrin of his two jealous brothers, who picked up lower offices. This made him governor of Azerbaijan province at a young age (seat of crown prince in Qajar tradition), which meant that it was he who had to ward off Russia's expansion into the Caucasus (Abbas-Mirza).

This engaged the crown prince for nearly three decades, and while he lost possession of these territories, it was not for lack of trying or ingenuity. There was simply too much lacuna to fill in such a short space of time (Peter the Great and Catherine the Great had given the Russians a huge head start).

The crown prince sent missions to Europe to bring back the latest military knowledge, engaged in active diplomacy (dropped the French as Napoleon signed off a deal with Russia, at Tilsit, so he and his father struck a deal with the British to receive military support), and he under-took various modernization initiatives. Technical progress by "regression method" was one of his innovations (i.e., as the troops needed gear, arms, and ammunition, he would acquire techniques needed to fabricate their constituent parts, assembled to the end-product, then in a next phase the building blocks of the building blocks, and so on toward primary raw materials, mastering the whole chain in backward sequence).

It was a loss for Persia that the crown prince never became a monarch as he predeceased his father, though the father also passed away within a year. The baton passed on to the crown prince's eldest son, as a sign of esteem held for the deceased crown prince, fixing the third monarch of Qajar. This monarch gave up on the Caucasus, now in full Russian control, and tried to reassert Persia's influence in Afghanistan, which brought him into conflict with the British. He had the beginner's luck, but at the end was unsuccessful; he lost Herat.

His son, the fourth monarch, attempted the same (to regain Herat), with the same result, then played with triangulation (engaging with Prussia, as well as overtures toward America), but he did not succeed. Thereby, he came to accept the Anglo-Russian dominance of Persia and only attempted to keep some sort of parity between their meddling in Persian affairs. The result was that as the world was moving forward, Persia was moving backward, for his rule lasted nearly fifty years (r. 1848–1896). This came to an end when a radical nationalist assassinated the king (with the motto *"the fish rots from the head not the tail"*).[25]

His son, the fifth monarch, conceded to pressure by moderate nationalists and made Persia into a constitutional monarchy (1906); but he passed away shortly thereafter. The latter's son, the sixth monarch, was brought up by the Russians from childhood (personal tutor), was fond of wearing Cossack clothing, and tried to restore the rule of absolute monarchy. He bombarded the *Majles*, but was brought down by the nationalists, who put his son on the throne, Ahmad Shah (r. 1909–1925), the seventh and last Qajar monarch.

Ahmad Shah was only eleven when he was put on the throne, and while Persia's national life was still dominated by Anglo-Russian hegemony, this situation changed dramatically during the course of the Great War. The Russians dropped out due to the revolution and civil war at home, and the British had not yet filled the vacuum, so for a brief period Persia enjoyed relative openness of political atmosphere, a *Majles* with political parties and a free press.

But Ahmad Shah was not deemed solid enough and cooperative enough to be the dyke against Bolshevik power. He refused to introduce the Anglo-Persian Agreement to the *Majles*, concocted by Lord Curzon and Sir Percy Cox, and Persia's Anglophile Prime Minister (Vosough-ol-Doleh), intending to make Persia into a British protectorate. This led to the emergence of Reza Khan on Persia's political stage, demise of Qajar, and the birth of Pahlavi (February 16, 1921 was the last visit of General Ironside with Ahmad Shah, February 18 the putschists left Qazvin for Tehran, four-day march with a pause in Karaj, and February 21 they arrived in Tehran).[26]

Though Reza Khan had wanted to make this a shock therapy, he quickly learned that that was impossible. He threw away his republic idea (borrowed from Turkey), as the high clergy threatened him with a *fatwa*. So, he reached a truce with the high clergy, married into Qajar nobility (his last three wives were of Qajar elite, the first three without stature, corresponding to his period of obscurity). He then picked up support from certain Qajar-era statesmen and carried on the programs that they had begun, but had not been able to push forward due to the war and its aftermath. Under Qajar there had been initiatives in the areas of law enforcement (Gendarmerie, 1910, created by the Swedes, in part to collect taxes from regional warlords), finance (recruiting controllers from America, Morgan-Shuster and Millspaugh), education (development of primary and secondary schooling, post-secondary institutes of political science, law, and medicine, some as offshoots of *Dar-ol-Fonun*), and secularization of judiciary (begun by Moshir and Dadgar, long before Davar). Reza Shah carried forth these programs, notably he merged the Gendarmerie with the Cossacks to create the nucleus of Persia's police and army.

There was therefore a good deal of branding in the Pahlavi reign, as there were more elements of continuity than discontinuity with the Qajar. There was a lot of Qajar in Pahlavi, far more than architectural relics and heritage. Reza Shah managed to carry out the implementation of this Qajar-initiated program by adherence to the principle of extreme precaution, which had serious ramifications in both domestic domain and foreign affairs, and this led to enormous dividends for the Pahlavi clan, accrued over a period of two decades.

The trial of the "fifty-three" was perfectly in line with this principle, and if it had one goal, it was to reinforce the image that Persia is in sync with the rest of the world; that is, the western world, anti-labor (all words connoting labor were banned in official media—*kargar, ranjbar, faa'le*) and certainly anti-communist, fully in control of its house. Germany, England, and the rest of the occident can count on the clairvoyance and competence of its leadership.[27]

All this stemmed from a profound fear of Russia, under whose tutelage he had toiled for several decades as a Cossack. He decided that if it had to be one or the other, it would be better to bet on Hitler than on Chamberlain, or whoever might succeed Chamberlain (he had dealt earlier with MacDonald and Baldwin), as a poisoned arrow pointing to Russia. Best, of course, was having both Germany and England on his side.

Taghi Erani was one element in the ensemble of symbolic acts executed by Reza Shah to convey this message of affinity with the Third Reich—the country's name change, bolstering Aryanism in official and semiofficial media (*Iran'e Bastan*, Ancient Iran, journal backed by Goebbels), anti-Arabism as Persia's version of anti-Semitism (an Academy set up to de-Arabize the language), swastikas in emblematic sites, scouts organization—a relationship buttressed by escalating trade and commercial transactions subscribing to the overarching strategy of collective autarky.

As for Taghi Erani—here was someone who was a nationalist, left-wing, independent, anti-racialist, internationalist, and pacifist; a Persian variant of Karl Liebknecht, sentenced to doom by the Pahlavi order, as in Pahlavi Persia there was no room for such misfits. This was the kind of language the Führer could understand.

As a further message woven into this act—see what happens to those that contemplate missteps—this part broadcast to the domestic audience. There was no harm to remind the nation, every once in a while, what awaits those who contemplate ill-thoughts. The Pahlavi order had to go on, and go on forever (resonating with the Führer's "we must build, and build for eternity").

* * *

The venue for the trial was a Qajar edifice: the ceiling, depiction of a battle between the Persians and the Russians, in which Crown Prince Abbas-Mirza figured prominently. The hall had raised sections above the ground, which in Qajar parlance is known as *Shah-Neshin*, where the Shah would take his seat. From this vantage point, he could survey the entire place, filled with his concubines, those handpicked by the eunuchs, as well as musicians, dancers, sommeliers, poets, jesters and guests.

Now *Shah-Neshin* (three of them) had become the seat of Reza Shah's judicial panel (judge and deputy, prosecutor and deputy, scribes and secretaries, all robed in black with a white margin). Below was the accused with their lawyers, not handcuffed but heavily guarded; behind them was the audience. The hall was filled with police agents in civilian clothing, personnel from concerned ministries, special guests (including the Police Chief and his card-playing buddies), handpicked journalists. No families, nor international observers or embassy staff were present. The British counsel's postmortem on the sentencing was based on journal reports. Stiff-armed security guards surrounded the hall: a show trial *par excellence*.[28]

Of the defending attorneys only two made remarks that challenged the charges frontally. One said that reading Marx's *Capital* is no crime, it is even recommended in western universities. And by the way, the person you refer to (Mr. Deputy Prosecutor) is not Friedrich English, but

Friedrich Engels. Another said the emphasis on creed in this trial is so pronounced that one naturally thinks of one's neighbor, but the time for such battles is long gone, is it not, pointing to the ceiling. There was an eerie silence, and he quickly moved on to other arguments. Both were summoned afterwards and became extremely circumspect (the implicit threat being that "fifty-three" can become "fifty-five," with the stroke of a pen).[29]

The other defending attorneys were mostly defending themselves. Explaining that they had not taken on these cases, but they were instructed to do so, and admitting that their defendants may have done wrongful deeds but did not have malicious intent. It was very pedestrian in tone. This person has done wrong but out of ignorance, please forgive him. Let us not worry about law and legality. A well-known attorney (Kasravi) said that one gets the impression that those accused were merely playing with the party idea rather than actually creating a party (*hezb-bazi* not *hezb-sazi*), somewhat as children play pretend. There is no harm in dismissing the case, a slap on the back of the hand is all that is needed.[30]

The defendants gave a variety of statements in self-defense. The Grenoble-educated lawyer used a highly legalistic tone to invalidate the charges. One of the accused who had been a bystander (his address had been used as a courier hub without him suspecting its purpose) said that his job is to repair people's shoes, not ask them their creed. But next time, he will make sure to do that, to not serve anyone of ill-faith. Another insisted that his precise response to the question be recorded, not what the court reporter read back to him; but the judge said it does not really matter (as if the sentences had already been cut out). Some pleaded remorse and asked for forgiveness. Reza Shah's son, the crown prince, and an Egyptian princess had become fiancées, and they were poised to tie the knot rather soon (March 1939, the trial November 1938), rumors circulating that a royal pardon was in the offing. Some of those accused were of reputable families, and the families had been doing quite a bit of lobbying outside to have their dear ones freed. They would share such news in their weekly visitations to lift morale.[31]

The trial lasted from November 2–12, 1938, November 13 closing arguments, November 15 sentencing. Taghi Erani's defense was on November 12, on the last day, for they did not want a spirited defense to arouse others into a combative mood. Two were acquitted (after eighteen months in captivity), and the rest got anywhere from three to ten years (this was the bracket defined in the anti-socialist law). Ten of the accused got ten-year sentences, as dictated by the Minister of Justice and Chief of Police. The Chief Prosecutor had recommended the maximum sentence for only three, and this after reviewing the interrogation reports given to him by prison authorities. The Minister of Justice and the Chief of Police considered that this was too lenient. There was the expectation that if the sentencing were not harsh enough, this would aggravate the king. For high-profile individuals (which in this case turned out to be only Taghi Erani), the sentencing did not matter, for they were sure to be eliminated in prison. Indeed, none of Reza Shah's political victims were put on trial, much less sentenced to death, except those tried in military tribunals (Puladin and Jahansuz), or in civilian courts without political charge (Teymurtash, embezzlement).[32]

Taghi Erani gave the performance of his life, reported to have lasted four hours. Various interruptions were attempted by the prosecutor, his deputy, and deputy judge, but the judge had allowed it to proceed. There are two versions of his defense plea (thirty pages each), not identical but highly similar, yet both are incomplete, as his defense had a general part and a specific part, the latter going into the details of the allegations and his refutations. This part is missing. And it went missing after the fall of Reza Shah.[33]

Documents that have been declassified reveal that the Grenoble-educated lawyer, having resumed his post in the Ministry of Justice (later he became *Majles* deputy and Minister of Labor, while the young king was still shaky), had checked out the dossier of the "fifty-three." Repeated reminders were sent to him, and it was returned after a lapse of several years, much of it missing, notably parts related to the Russian-trained pilot. Most likely he had passed on the dossier to the Soviets (who occupied northern Persia, 1941–1946), sanitized it, and returned it to him, who returned it to the Ministry. In this process, the major part of Taghi Erani's defense was also lost, at least in the Persian archives.[34]

Within weeks after the fall of Reza Shah, political prisoners were freed. About ten of the "fifty-three" were among the founders of a new party called *Tudeh* (Masses). *Tudeh* did not hesitate to make political currency out of the name of Taghi Erani. They gave the name *Donya* to their party's journal, the name used by Taghi Erani for his journal. They indicated in their journal that Taghi Erani had been its founder. They advertised Taghi Erani as their party's spiritual leader, which was odd as they were subject to Soviet influence while Taghi Erani represented autonomy in political practice. The founders of this party included the Grenoble-educated lawyer, but not the pilot. He would join this party in two years, shortly after its official inauguration (groundwork began right after the abdication of Reza Shah). And he would rise not only in this party but also in another which seized political power in Azerbaijan, with the agenda to merge it with Soviet Azerbaijan. This was later resolved at great-power diplomatic level, in the context of departure of Allied forces from Persia (Soviets let go of their support, allowing the young Shah to march in victoriously, 1946). Two years after the fall of this insurgency, the pilot left Persia for Russia and later went to East Germany, where he passed away (Leipzig, 1970). In memoires of a Soviet agent, Ilya Dzhirkvelov, there is a whole section on him and how Stalin approved the operation to rescue the pilot out of Persia in 1948. But before that, during his two-year absence from Persia (1941–1943), he had gone to the Soviet Union, was sheltered by Baghirov (President of Soviet Azerbaijan), wrote a report on the "fifty-three" to Pavel Fitin (the NKVD deputy director), who communicated it to Dimitrov (head of Comintern), who reported to Stalin (however, there is no record of Dimitrov-Stalin correspondence on this affair, only transmittal letters of the subordinate reports, made accessible in mid-1990s, post-Gorbachev era, published in 2008).[35]

***

Reasonable conjecture as to how the pilot was rehabilitated in the Soviet apparatus, after the debacle of the "fifty-three," and what had motivated his extensive cooperation with the Reza Shah police is somewhat as follows—as an elite NKVD agent, he had to ensure his self-preservation, as well as preservation of strategic assets of the Soviet Union in Persia, and elimination of individuals who had lost utility or had become a nuisance.

He had served in the Air Force and had a prior judicial record; so, the main risk was referral to a military tribunal and execution. Although such act was not without risk for the government of Persia, for Russia could retaliate, there was precedence to this effect (an agent was executed in the wake of Agabekov defection). Execution was therefore a real possibility. Thus, the only solution was to cooperate but not divulge strategic assets. This included certain of his circle (which did fall into the group of the "fifty-three," notably Zia Alamuti or *Fattane*, but through confession of others), and more importantly, the moles he had implanted in the armed forces. (These remained intact, Colonel Siamak the only one known to us, the names of others we cannot be certain about; however, *Tudeh* came to have an embedded military organization in the 1940s, estimated at five hundred, some of whom were recruited in the 1930s, most likely those who led the revolt in Khorasan in early 1940s, colonel Azar and Eskandauni among them.)[36]

The pilot was presumably given assurances that he would be kept out of the picture if he cooperated fully. This was the intent of the authorities, who did not want to implicate the Soviet Union. (Although they would use this in a tactical sense to extract confessions, in accordance with the Persian saying—"*to threaten with death, so they settle for illness.*") There may even have been hints of a swap (notably with Persian communists held in the Soviet Union, such as the CRRP co-founder, Morteza Alavi, particularly of interest to Reza Shah, as he had caused the king's embarrassment in the German press during Weimar). But a swap did not materialize, either because the Russian half of the barter was already eliminated, or because a barter was not envisaged or investigated.

The pilot therefore became a close collaborator of the police. He named not only those in the Erani circle but also those that Erani had named as potential recruits for the future (the two had regular meetings from September 1935 to March 1937, twenty to thirty times, according to Taghi Erani). The pilot also elaborated on the assigned organizational roles and activities of these members, producing in effect a blueprint for the police. His collaboration with the police continued while the "fifty-three" were being interrogated. He would receive reports of confessions and guide the police in their further questioning. He also divulged the past history of Taghi Erani, from Berlin years, what he knew from his Comintern briefing (for Taghi Erani was extensively questioned on this, although it had been a closed case). The ultimate goal of the police was to demonstrate that Taghi Erani was the leader of a clandestine group, led by a left-wing political ideology, illegal in Persia, with the intent of toppling the government, and that he was also the person who had divulged everything to the police. This would be a psychological blow to anyone engaged in anti-government activism. To hammer the message that even those who pretended to be dissident leaders end up crumbling. Taghi Erani knew many of those arrested and could be shown to have been the first that was arrested. There was no one else of comparable importance in the group, the Berlin-educated medical doctor playing a distant second fiddler. The goal of the police coincided with the pilot's objectives, which had been to dispense with Erani once all useful information and links were extracted from him. Elimination was the best option.

The plan to keep the pilot in disguise, however, did not work, as the two prime characters, Taghi Erani and the Berlin-educated doctor (Mohammad Bahrami), categorically denied the existence of such an organization, despite the use of various coercion tactics by the police. The police interrogation then focused on whether they knew the pilot (implying that he is at large) and when this too was denied, they brought him out of darkness and created one-on-one confrontations. This caused certain facts to be admitted, but it also caused the pilot's case to become integrated with the rest. And later, because of a lapse in coordination between the police, the prison authorities, and the Ministry of Justice, the inmates learned that the real informer had been the pilot. This became evident in August 1938, fifteen months after the arrests, when their confession reports were read to them as a group, to not give them access to the files, and to save time in view of the impending trial. Though they still did not discover that the pilot had been picked up weeks before them, the dating scheme, fabricated by the police, showed that he was picked up two days *after* Taghi Erani, not one month before.[37]

After the trial, and for the rest of the time the pilot was in prison, three years and two months, the first half done in Tehran, the second in Bushehr, he kept a low profile, cooperated with others as much as he could, and aroused their pity as having shown weakness because he had been scared for his life, due to his prior history. No one suspected that he had cultivated a deeper plan. He was also cunning in his dealings with cell-mates, would share the money brought for him from outside, offer cigarettes, not show off his knowledge and experience, and blended into the landscape.

Taghi Erani, on the other hand, was made out to be the informer by the police, the pilot, and pilot's inner circle, something which was taken to be a certainty by the great majority of the group, until August 1938, at the collective report-reading session. After that, he was kept at the detention center, not *Qasr* prison, separate from the others (with the exception of three weeks). Thus, during the first half of a thirty-three-month period of captivity, he was considered to be a traitor.[38]

Upon release from prison, the pilot had gone to the Soviet Union to explain what had happened and the rationale for his collaboration with the police and the way he had engineered it. He had had three channels at his disposal—political, ideological, military: the first with his hierarchy in the Soviet Union to receive directives, the second via Taghi Erani, to cast a net wide enough to catch as many contenders as possible for recruiting (to scavenge intelligence and disseminate propaganda), and the third through his network of moles in Persia's armed forces. In effect, he had sacrificed a part of one branch, the part that had to be decapitated, as a tactical gambit, to preserve the strategic parts. Following this elaboration, he was rehabilitated by Soviet authorities.[39] Proof is the prominent position he occupied in the 1940s in the Soviet-backed *Tudeh* party and in the Soviet scheme to annex Azerbaijan. Indeed, when he returned to Persia in 1943, to assume a leading role in *Tudeh* and the provincial party in Azerbaijan, he stated that Soviet authorities had executed the old cadres of Persia's communists in exile in the Soviet Union, because they had been found to be guilty, having caused the discovery and the arrest of the "fifty-three." Thus, were executed, among others, former collaborators of Taghi Erani—the CRRP co-founder (M. Alavi), the intellectual who had been Taghi Erani's neighbor in his youth (Dehzad), and the other who worked closely with the latter and had offered a critique of Taghi Erani's *Psychology*, advising him to stay out of political activism (Ladbon).[40]

These activists had their own analysis of Persia's situation and requisite method of revolutionary work, at variance with the ultra-left stance of the sixth congress of the Comintern, and the ultra-right and wait-and-see approach of the seventh congress, the United Front dictate. They all had soft points which could become pretext for their elimination—one had had a girlfriend in Germany who turned out to be a police agent, another had overspent from the party's funds (so it was claimed), another had dabbled in bourgeois art (modern poetry), and so forth. When the Comintern decided that these individuals were not cut out to be agents and were "merely" activists, they were set aside for later decision (the CRRP co-founder was sent from Moscow to Tashkent to help with party work among those of Persian heritage, the other having conceded to overspending funds accepted to work in an automobile factory to be rehabilitated). At any rate, they were eliminated on drummed up charges of turning in the "fifty-three," while the pilot was cleared of wrongdoing and received full backing from Soviet authorities.

In essence, the old communists had been executed because they had outlived their usefulness, similar to what happened to Bulgarian and Polish communists in exile in the Soviet Union. With the relative openness of political atmosphere in Persia after the fall of Reza Shah, the risk that they would slip back into the country and divulge information that would be damaging to the Soviet Union was an unacceptable exposure. Only a professional agent could be entrusted to be there and act responsibly. Therefore, the pilot was dispatched back to Persia, while others were eliminated.

Nobody understood this turn of events better than the Grenoble-educated lawyer, who paved the way for the pilot, at least in connection with vandalizing the legal documentation of the "fifty-three." The lawyer had his heydays in the 1940s, bolstered by *Tudeh* (making it as *Majles* deputy and Minister of Labor), left in time not to be caught up in the 1953 *coup* (return of Shah to power in operation AJAX, after fleeing the country caused by wave of nationalism which brought about the premiership of Mossadegh, nationalization of oil, demise of APOC). The lawyer homesteaded in East Germany, where he passed away some years later (Leipzig, 1985). The two Berlin-educated medical doctors, with opposing character traits, also thrived in the 1940s, in *Tudeh* leadership (the one with bourgeois manner became Minister of Health). Both were caught after the *coup*, did some time in jail, and through intervention of influential figures and personal pleas were released. Both passed away in Tehran: the hardy character in 1957, the posh in early 1980s. As for the others, the majority stayed clear of politics after their release from prison; such was the case of the smuggler. The rest dabbled in politics, mostly in connection with *Tudeh*. Some of these broke off within a few years, creating splinter groups, which faded after the *coup*, as did the *Tudeh*, a party in exile with an unsuccessful bid to return after the 1979 revolution; *Tudeh* was annihilated in 1982. This was the other part of the aftermath.

NOTES

1. See Kapuscinski, *Le Shah*, pp. 43–44.
2. See Pasyan and Motazed, *Zendegi'e Reza Shah*, pp. 716–717, for memoires of Majid Movaghar, TIRR official and editor of *Iran* newspaper, about the Basra buildup.
3. Hitler's attack on Russia was operation Barbarossa (June 22, 1941). See Majd, *Great Britain and Reza Shah*, pp. 131, 264–265, for haggling over oil royalties in the midst of the war; see also Afshar, *Zendegi'e Toufani*, for Taghizadeh's account, pp. 242–243. For expiration of Reza Shah's usefulness, his end, see Elwell-Sutton, *Modern Iran*, pp. 174–194. For his health and opium habits, see Pasyan and Motazed, *Zendegi'e Reza Shah*, pp. 247–248, 745–760 (latter is allegedly testimony of his son, Abdol-Reza Pahlavi); his bad teeth and ulcers, pp. 772–773.
4. Author's father (Asadollah Jalali-Yazdi, 1920–2012) was dispatched from Mash'had to Sarakhs near Turkmenistan border; they resisted three days against Russian onslaught, before dispersing, and later having to turn in their arms when Khorasan went under Red Army occupation (1941–1946). His certificate of completion of service states the reason for his discharge was the "events of 3-Shahrivar-1320." Shortage of grain and bread was a major preoccupation. The Russians built a grain silo in Khorasan, where he worked as a construction worker, picking up some Russian in the process. For list of towns occupied by Russian and British forces (35 Russian, 18 British, within a week), see Pasyan and Motazed, *Zendegi'e Reza Shah*, p. 734. American forces arrived in December 1941, after Pearl Harbor triggered US entry into the war.
5. See account by Jalal Abde, senior magistrate, in Golbon and Sharifi, *Mohakemeye Mohakemegaran*, pp. 409–415. Those charged were— Chief of Police (Rokn-ol-Din Mokhtari), Chief of *Qasr* prison (Hossein Niroomand), his predecessor (Mostafa Rasekh), chief executioner (Pezeshk Ahmadi). Unclear whether chief of detention center (Sartipzadeh) was among them. Other perpetrators are listed in Fami-Tafreshi, *Polis'e Khafiye' Iran*, pp. 232–295.
6. It was Mohammad-Hossein Ayrom; see Amanat for his bio. See Afshar, *Zendegi'e Toufani*, pp. 228–230, for Taghizadeh's account of how Ayrom operated. See Fami-Tafreshi, *Polis'e Khafiy'e Iran*, pp. 139–159, on the Ayrom era (1931–1934), and Motazed, *Polis'e Siyasi*, for detailed account of his methods.
7. See Golbon and Sharifi, *Mohakemeye Mohakemegaran*; present author had correspondence with the now late Mr. Mohammad Golbon, facilitated by Mr. Ahmad Bashiri (literary figure, editor of Divan of Abol-Ghassem Lahuti).

8. See Motazed, *Polis'e Siyasi*, p. 387, and Fami-Tafreshi, *Polis'e Khafiy'e Iran*, p. 293, and Bozorg Alavi, *Panjah'o Se Nafar*, p. 223, on Mokhtari's character and his musical career; allusion to card-playing is by Khame'i, in *Panjah Nafar va Se Nafar.*

9. For getaway attempt of executioners, see Golbon and Sharifi, *Mohakemeye Mohakemegaran*, pp. 409–415; for Mokhtari on national radio, see Motazed, *Polis'e Siyasi*, pp. 378–392; on Matin-Daftari threats, see testimony of Maître Shahidzadeh in Boroujerdi, *Erani Faratar az Marx*, p. 703, also testimony of Tabari and Alavi, *Zendegi'nameh Dr. Erani*, p. 41, and Khame'i, *Panjah Nafar va Se Nafar*, p. 169; on fate of Matin-Daftari, see Abrahamian, *Iran Between Two Revolutions*, p. 241. He was not held for very long by the Allies and later became a senator.

10. Expulsion of a left-wing student happened in May 1933, see Jenkins, *Hjalmar Schacht, Reza Shah, and Germany's Presence in Iran*, p. XXXV (footnote 50), no name given. Mohammad Bahrami was KPD member and in Germany when the Nazis came to power; he left in 1934.

11. My account of early life of Reza Shah is based on memoires of General Hamzeh Pasyan (1849–1924) who was in the PCB. His handwritten notes were transmitted to his son Najaf-Gholi Pasyan, the cousin of the famous Mohammad-Taghi Pasyan, and is narrated in Pasyan and Motazed, *Zendegi'e Reza Shah*, pp. 43, 49–50.

12. Reza Shah and misappropriation of oil revenue; see report of Louis G. Dreyfus (American Minister), in Majd, *Great Britain and Reza Shah*, pp. 267–270.

13. See Kapuscinski, *Le Shah*, p. 44, on Sir Reader Bullard's stern words to Reza Shah, in effect to abdicate; for the case of his son, Mohammad-Reza Shah, see General Robert Huyser's account in *Mission to Tehran.*

14. See Valizadeh, *Etehad'e Jamahir'e Shoravi va Reza Shah*, pp. 203–206, upheavals in Russian-Persian relations.

15. See Brook-Shepherd, *The Storm Petrels*, pp. 5–24, on Bajanov defection. For Soviet infiltration in Persia, see Agabekov, *O.G.P.U.*, pp. 73–85, 103–146, 171–178, and Valizadeh, *Etehad'e Jamahir'e Shoravi va Reza Shah*, pp. 206–213. See memoires of Minister of Transport (Dr. Mohammad Sajjadi) at the time of Anglo-Soviet invasion (August 1941) and Reza Shah's regular calls to him to know how far the Russians had advanced, in Pasyan and Motazed, *Zendegi'e Reza Shah*, pp. 271–272.

16. Reza Shah life in exile; see Pasyan and Motazed, *Zendegi'e Reza Shah*, pp. 741–781, account attributed to his son (Abdol-Reza Pahlavi). Periods—Mauritius Island (October 1941–March 1942), Durban (April–May 1942), Johannesburg (May 1942–July 1944).

17. For Reich-Pahlavi relations, see Elwell-Sutton, *Modern Iran*, pp. 164–168, Valizadeh, *Etehad'e Jamahir'e Shoravi va Reza Shah*, pp. 243–260,

and Jenkins, *Hjalmar Schacht, Reza Shah, and Germany's Presence in Iran*. On corroboration of Russia and Great Britain with German presence in Persia, see Valizadeh, pp. 251–260.

18. Reza Shah, disdainful view of Nader Shah ("that brigand"), see report of American chargé d'affaires, J. Rives Childs in Majd, *Great Britain and Reza Shah*, p. 202.

19. Visit of Hjalmar Schacht to Persia (November 20–24, 1936); see Jenkins, *Hjalmar Schacht, Reza Shah, and Germany's Presence in Iran*; also Ahmadi, *Khaterat'e Bozorg Alavi*, pp. 146–147 (anecdote of Alavi's exchange with Schacht who visited the School of Industry, formerly German School).

20. See Boroujerdi, *Erani Faratar az Marx*, p. 342, on Erani's initiative to phase out reliance on German experts at the Ministry of Industry.

21. See Ahmadi, *Khaterat'e Bozorg'e Alavi*, p. 215, on premeditated nature of eliminating Erani; "they really had the goal to kill him."

22. See Momeni, *Parvandeh*, pp. 88–90, 294–297, for interrogations and house searches to pinpoint communist literature in Erani's possession. Interrogations were structured to prove that this group is a political entity (*ferq'e*), that it is communist (*maram'e eshteraki*), and that he is the ringleader (*pishva*, leeder). In his defense plea, Erani stated categorically that Persia's constitution guarantees freedom of thought and expression, does not preclude socialist or communist ideas, and that the 1931 law banning these is unconstitutional (*ghanoon'e siah*, "black law"). He also cites constitutional governments that freely allow such political expressions in England, France, USA, and Switzerland; see Momeni, *Parvandeh*, pp. 267–268. In his interrogations, he refutes the charge that this group is a political party (p. 69) and the leadership charge ("how could I be the leader when I did not even know the identity of Kam-Bakhsh and Shureshian," pp. 276–277). He also highlighted that veteran communists considered his views unorthodox (p. 238). Also see Boroujerdi, *Erani Faratar az Marx*, pp. 328–330.

23. See Elwell-Sutton, *Modern Iran*, p. 168, also Jenkins, *Hjalmar Schacht, Reza Shah, and Germany's Presence in Iran*. Baldur von Schirach arrived in early December 1937. He had stops in Turkey, Iraq, and Syria, but it was in Iran that he was most interested, as he explained to Hitler—"The main goal of this trip, Iran, is of special interest to me because in a relatively short period of time a youth organization was established with a national importance that is considered to be very high" (quoted in *Nazi Germany and the Arab World*, Francis Nicosia, p. 98). He was married to Henriette Hoffmann, the daughter of Heinrich Hoffmann, Hitler's photographer and member of his inner circle.

24. Venue for the trial of "fifty-three," *Haramsara* of Fat'h-Ali Shah Qajar, depicting in its ceiling the Persian–Russian battles over Caucasus (1803–1828). It was at *Bab'e Homayoon*, near *Golistan* palace. See Khame'i, *Panjah Nafar va Se Nafar*, pp. 163–167.

25. The radical nationalist was Mirza-Reza Kermani (c. 1850–1896).

26. See Ironside, *High Road to Command*, pp. 164–168.

27. On banning words connoting "labor," see Khame'i, *Panjah Nafar va Se Nafar*, p. 17. Also see Erani, *Defaiyat* (in Tabari and Alavi, *Zendegi'nameh Dr. Erani*, pp. 65–66): "Isn't Persian society nine-tenth laboring class; the ban shows not only you despise them but you also fear them…".

28. British minister in Persia, Mr. Seymour, characterized the sentencing in this case as having the goal to frighten the nation, see FO 371/21890. Also see Boroujerdi, *Erani Faratar az Marx*, pp. 348, 699–703. Latter is testimony of Maître Shahidzadeh on how this was a police operation in the guise of a legal operation.

29. Two attorneys who showed courage, threatened by Minister of Justice (Ahmad Matin-Daftari); these were Dr. Aghayan (forename not indicated, he was attorney for Dr. Bahrami) and Ghavam-ol-Din Majidi; see Boroujerdi, *Erani Faratar az Marx*, pp. 655, 687–688, based on testimony of Nosratollah Jahan-Shahlu-Afshar, one of the "fifty-three." See pp. 359–360 for testimony of Jalal Abde, senior judge, who had to investigate Dr. Aghayan's "transgression" and report it to the Minister of Justice. For description of defense given by these two attorneys, also see Khame'i, *Panjah Nafar va Se Nafar*, pp. 168–169.

30. Other attorneys were—Seyyed-Hashem Vakil (for Taghi Erani), Arsalan Khalatbari (for Kam-Bakhsh), Kaumkar (for Radmanesh and Khame'i), Ahmad Kasravi, Amidi Nuri, Manshoor, and Nonahal. See Khame'i, *Panjah Nafar va Se Nafar*, pp. 165–171.

31. Ibid., pp. 165–171; also p. 228, for possible amnesty due to crown prince's forthcoming marriage.

32. See Boroujerdi, *Erani Faratar az Marx*, pp. 355–357, for account attributed to the judge, that all sentencing was dictated by Chief of Police and Minister of Justice; when he objected that he could not give ten years to Iraj Eskandari, an employee of the ministry, he was told then give it to someone else, but the number of those getting the maximum sentence of ten years cannot change (ten people); see pp. 431–442, for details of the sentencing of the entire group.

33. See Erani, *Defaiyat*, two versions reproduced in Boroujerdi, *Erani Faratar az Marx*, and in Tabari and Alavi, *Zendegi'nameh Dr. Erani*. Memos exist from Ministry of Justice to Iraj Eskandari, employee of the ministry, who had checked out documents of the trial; see Boroujerdi,

Erani Faratar az Marx, p. 500, for memo with date of 11-Azar-1321 (December 2, 1942), and p. 505, memo of 15-Aban-1328 (November 6, 1949), indicating it is still missing, and intervening notices. See also Momeni, *Dar Khalvat'e Doost*, p. 98, where Bozorg Alavi attributes disappearance of the judicial file of "fifty-three" to Kam-Bakhsh to cover up his "mess."

34. Ibid.
35. See Dzhirkvelov, *Secret Servant*, pp. 72–74. For Kam-Bakhsh's report to Comintern (to Pavel Fitin, in turn reporting to Dimitrov, head of Comintern), see Chaqueri, *Taghi Erani dar Ayeneye Tarikh*, pp. 259, 263. For his "running away to Soviet Union" after release from prison and his collaboration with the police while in prison, see Ahmadi, *Khaterat'e Bozorg Alavi*, pp. 161, 219–221; also see Katouzian, *Khaterat'e Khalil Maleki*, pp. 282–283. For details of Kam-Bakhsh collaboration with the police, see Momeni, *Parvandeh*, pp. 31–44.
36. For execution of suspected Soviet agent (Mohammad-Ali Mirza Khaghani), see Valizadeh, *Etehad'e Jamahir'e Shoravi va Reza Shah*, pp. 66–67. For Kam-Bakhsh's prior offense in Persia's Air Force, see Momeni, *Parvandeh*, pp. 38–39. For his admission and justification for collaborating with the police, see Boroujerdi, *Erani Faratar az Marx*, pp. 270–275; this is based on testimony of Babak Amir-Khosravi present in the Fourth Plenum of Tudeh (June 1957, Moscow), where questioning was done on this issue and Kam-Bakhsh's reasoning was dismissed by the majority faction of Tudeh, led by Radmanesh and Eskandari. In this event Eskandari called Kam-Bakhsh "Qatel'e Erani" ("Erani murderer"). See Boroujerdi, Erani *Faratar az Marx*, pp. 272–273, account by Eskandari: "Erani said, I have also written many books, but I would be totally incapable of writing a book as concise as he has, referring to Kam-Bakhsh's confessions, in such a short period of time."
37. See Momeni, *Parvandeh*, pp. 34–38, for police's plan to not reveal Kam-Bakhsh and misfiring of the plan. Soviets had experience in barter of key figures. Matyas Rakosi was in a Hungarian jail, but released to the Soviets in 1940 (in exchange for historical artifacts that had been stolen by the Soviets). He later returned to Hungry and led its communist party (1945–1956) with pro-Soviet policies. It is unclear whether barter was considered in the case of Kam-Bakhsh, but Morteza Alavi was one figure of great interest to Reza Shah, due to the Weimar court case that he had won against the government of Persia. Morteza Alavi was in the Soviet Union since 1933. He was executed in 1942.

38. Taghi Erani, prison periods: May–December 1937 at detention center (*Zendan'e Movaghat*); January 1938, two weeks at *Qasr* prison; mid-January–August 1938 at detention center; most of September 1938, at *Qasr* prison; October 1938 at detention center; November 1938, two weeks at *Qasr* prison (during the trial); mid-November 1938–February 1940 at detention center, ending with his engineered death.

39. Kam-Bakhsh left for the Soviet Union immediately after his release, staying there two years (1941–1943). Most of this time he worked at the radio station in Baku, protected by Mir Jafar Baghirov, the strongman of Soviet Azerbaijan (head of the Republic). When he was cleared by Comintern, he returned to Persia with strong Soviet backing; he was in Persia during 1943–1948 (key figure in *Tudeh* party and *Democrat* party of Azerbaijan, latter led by Pishevari), and thereafter lived in Moscow and Leipzig (1948–1970), where he passed away. See Ahmadi, *Khaterat'e Bozorg Alavi*, p. 161, stating that Kam-Bakhsh made arrangements with the Soviets (1941–1943), came back and took control of the "ship's helm" (*sokkan'e keshti*) of *Tudeh* party.

40. Persia's veteran communists residing in the Soviet Union, eliminated during reign of Stalin include–Sultanzadeh, Seyfi, Sharghi, Zarreh, Nikbin, and Akhundzadeh. Also, the three that intersected with Taghi Erani—Morteza Alavi (who had had a girlfriend back in Berlin, alleged German undercover agent), Abol-Hossein Dehzad (charged with overspending party funds), and Ladbon Esfandiari (experimented with so-called bourgeois poetry).

CHAPTER 15

The End (1937–1940)

Taghi Erani was arrested on May 8, 1937 and met his induced death on
February 4, 1940, short of his thirty-eighth birthday (March 21, 1902).
His captivity lasted nearly thirty-three months. He spent virtually all of
this time at the detention center in the capital (*Zendan'e Movaghat*), not
at *Qasr* prison, which was in the outskirts of the city, but in the center of
town, at the police headquarters, the *Shahr'bani*. This was the opposite
to the rest of the "fifty-three," who spent few months at the detention
center and the rest at *Qasr*. They were all freed after the fall of Reza Shah.

Qasr was where one could interact with many others—veteran political prisoners, trade-union activists, nomadic chieftains, military officers,
fallen politicians, journalists, administrators, and some foreigners
(Ukrainian-Germans fleeing Stalin's collectivization of agriculture). All of
these were related to political, security and trespassing charges. As well
there were regular prisoners, offenders related to robbery, battery, homi-
cide, and variety of criminal activity.

Inmates at *Qasr* were held in different wards depending on their
profile, but there was still a lot of diversity within each ward and occasional cross-ward transfers. There was also a ward for women. There
were nine wards in total, six large, three small, the small ones the size
of a large ward if put together. The total number of inmates was about
two thousand. Each large-sized ward was comprised of about three hundred inmates. Each ward was separated into corridors, and each corridor into a number of cells. The great majority was for grouped inmates;
five or ten in a cell, and one with a hundred or more, was fifteen meters

© The Author(s) 2019
Y. Jalali, *Taghi Erani, a Polymath in Interwar Berlin*,
https://doi.org/10.1007/978-3-319-97837-6_15

by eight, for common criminals, taking turns to sleep. There were cells for solitary confinement, typically two-and-half meters by one-and-half. Each ward had its own courtyard used for "breathing time," about half an hour daily, taken on a rotation scheme. *Qasr* is where there was visiting rights for the families, every Tuesday afternoon, with short time slots. There were four watchtowers overlooking the compound and a caserne nearby.[1]

The detention center, however, was different; for one thing, there were no visitation rights. It also had wards, four in total, overlooking a common courtyard, triangular in shape. One edge faced wards one and three; another, wards two and four, and the third faced the rest of the police headquarters. Each ward was a mix of solitary and group cells, the latter with view of courtyard. The compound could hold a few hundred inmates at most. Its "clinic" is where well-known political murders took place (Hejazi, trade unionist, Teymurtash, Court Minister, Sardar Asad, Bakhtiyari baron and Minister of War, Farokhi-Yazdi, the poet–journalist and deputy). All these were before Taghi Erani, though the last one was only four months before his death. The clinic was where doctor death (Pezeshk Ahmadi) had all that he needed to execute tasks assigned to him by the chief of the detention center or the chief of *Qasr*, or their superior, the Chief of Police.

The detention center was at the heart of *Shahr'bani*, the police headquarters in the capital, near Artillery Square (*Toop-Khaneh*), a few kilometers from *Golistan* palace, near *Dar-ol-Fonun*, where Taghi Erani had gone to school in his youth. The compound was built by the Swedes (c. 1910), who founded modern policing in Persia. The Qajar picked countries that had no vested interests in Persia to procure technical assistance: Belgians established the customs administration, Americans finance, Swedes law enforcement; a mix of Europeans, mostly German-speaking, staffed *Dar-ol-Fonun*, and there were a number of European technocrats in the postal service, telecommunications, transportation, and health, a modest Meiji.

The detention center was a two-story structure, cells at ground-level, above-ground bureaus for administration, interrogation, and coercion. Swedish cuffs and padlocks were still in use. One method that was commonly used consisted of clasping the left hand reaching over the left shoulder with the right hand reaching from right side of the lower back; the tension could be adjusted, causing the dilation of the chest bones over time. Weights attached to the testicles were another form of torture, as was a variety of cuffs for wrists and ankles. Physical assault, threats to lock up family members, confinement in solitary cells that were damp, infested with lice and insects, without a bed, with an open sewage, were other means of punishment.

Taghi Erani was moved from the detention center to *Qasr* prison on three occasions. First was right after formal charges were brought against the group (December 1937); it lasted about two weeks (January 1938). The second time was in anticipation of the trial, with the intent to keep him there till trial was over, but it came to an abrupt end, after only three weeks (September 1938). Before this second transfer, he attended a one-day session at *Qasr*, where excerpts of interrogation reports were read out to the group, and another day for assignment of defense attorneys. The third transfer was during the trial (November 1938), which was held in a Qajar monument in the capital. Inmates were brought in from *Qasr* in the morning and taken back to *Qasr* in the afternoon. There were seven or eight buses used for this purpose, as there were no more than seven or eight inmates in each, with double that number in guards. The same procedure was used when transferring inmates between the detention center and *Qasr*. This third episode lasted two weeks.

Thus, Taghi Erani spent a total of seven weeks at *Qasr* (hence at most seven family visits) and thirty-one months at the detention center. At the latter place, he was in solitary confinement the whole time (with the exception of three days, end-December 1937, before being taken to *Qasr* for the first time, with the rest of the "fifty-three"). He was initially in a cell without toilet, so he would be taken out for his needs, but later he was placed in a cell with a toilet, condemned to confinement for days on end. But at *Qasr*, he was in a group cell in ward two, which was connected to wards four and seven, the three being the small wards at *Qasr*, with a common courtyard. All members of the "fifty-three" were in these three interconnected wards, as well as veteran political prisoners, and a large contingent of activists from Rasht (the group of "twenty-three" connected with Ismail Foroohid), and others who were mixed in, not necessarily political prisoners (one journalist). All told, there were about two hundred inmates in these three wards.

The detention center had an agitated atmosphere. It kept those who had just been brought in and were to be "processed," or those who had been brought back from *Qasr*, as they had misbehaved. It was the place where harsh punishments were meted out. *Qasr*, however, was more settled and subdued. It was larger, more varied, with contact to the out-side world. Those who had the means could use cash to get a variety of things, or receive packages from visitors—soap, medicine, food, journals, books; opium could be obtained through *Qasr* dispensary (it was made into a dilute solution, called *ab'e-taryak*). However, the guards could come and search for these same items they had smuggled in for a fee, or turned a blind eye to, and take them away. It was somewhat arbitrary.

This convenience factor had its ebb and flow. For instance, when the head of *Qasr* retired (September 1938), and a new director was appointed, tough measures were implemented to set the tone. Personal heaters were removed, and it became impossible to get reading mate-rial. Moscow show trials, civil war in Spain, fall of Czechoslovakia, the crown prince's wedding, and French journals mocking the Shah (*chat* or cat), jab for his pro-German postures, were the hottest newsbytes traded amongst prisoners, when they had access to news.

Some of the items brought in seem like luxuries, but it was widely believed that they were vital to survival. For instance, if one depended solely on prison food and had no means of heating it (*aladin*, kerosene burner), it was a sure recipe for all sorts of illnesses. Some level of brib-ing went on at the detention center as well; no visitors were allowed, but families could send things (clothing, food, medicine), and depending on the whims of those around, it could get delivered or stolen.

Some of the prison workers were prisoners with little means, who made pocket money doing chores of the regular employees as gofers. They tended to be from the provinces, little towns and tribes. They would get money from the prisoners for a variety of favors, bringing in cigarettes or a journal, and handed over most of that to the regular staff, keeping small change. They also had to report on anything suspicious they saw. Inmates who had no money suffered greatly. Some made hand-iworks for petty cash (visages of famous people woven with straw and sticks), bongs made out of eggshells, chess pieces out of cork, cigarettes were good enough as means of payment. Winters were harsh as the place was not heated. Prisoners had a mat and cover, unless it was taken away as punishment, inflicted at times on those in solitary.

***

The year 1937, May to December, from arrest to indictment, the main events experienced by Taghi Erani were as follows—six interro-gations conducted over a seven-week period, with increasing tension, the last ones not in his handwriting, due to refusal to sign and physical assault. He stressed that the pilot was playing with people's lives and the smuggler was a thug and untrustworthy. He conceded knowing the CRRP co-founder in Berlin (Morteza Alavi) and having read books considered Marxist or Communist and having had discussions of a philosophical nature with various people. But he categorically denied the charge of creating a party or attempting to topple the government or being subordinate of a foreign power.[2]

He admitted to knowing the pilot and having had discussions about various people who had progressive inclinations. But he was not involved in organizational matters. He also admitted to knowing the person who had introduced the pilot, who had insisted that his journal be turned into a Comintern mouthpiece, which he had refused, although he did make a financial contribution to this person and did meet him in Moscow (Kamran), on his return journey from Berlin to Persia. He also admitted to having met the one who had been his neighbor in his youth (Dehzad) and his associate (Ladbon), both in Tehran.[3]

These themes were explored in the interrogations, to see what links may have existed in the context of the Comintern or Soviet intelligence. But in official charges and the sentencing, not a word was uttered about these matters. The seven-month period from arrest to declaration of formal charges was also a period of debate by the authorities as to how to deal with this situation. Lock up the people without declaration, make a statement implicating the Soviets, or declare the arrest and couch it as a domestic affair; latter is what they opted for at the end, though news of the arrest did spread very quickly.

There were also five requisitions made at his home, investigators were accompanied by a translator, rummaging through his books, letters, and paperwork in his office, and taking away parts of these in a sealed suitcase, for which his mother and youngest sister had to sign off.[4]

At the detention center, his cell was near the ward's toilets. This allowed him to utter phrases in French or German when those whom he knew would pass by. He also managed to leave a note for one of the "fifty-three" under the toilet receptacle used for self-washing. It was on cigarette paper (though he was not a smoker) and was a will of sorts, stating that they may exterminate me, but they will free you (addressing the novelist who collaborated with Taghi Erani on the journal, Bozorg Alavi, and similar message he gave orally to another as well, Jahan-Shahlu-Afshar), make sure to visit my mother and make sure my little sister gets a chance to pursue her studies. A second time he left a note, he was caught, and this caused great hardship. The note said "the police has turned savage, cooperation will lead to more savagery, we must resist." A teenager who worked as a cleaner reported this, and much suffering ensured. (The police's line was that in prison "We" does not exist, only "I" exists.)[5]

This caused his transfer to a cell with a toilet, so he was deprived of leaving his cell (September 1937). This was cell no. 28 in ward three. A mason was brought in to re-plaster the wall removing etchings previous prisoners had left. He used this opportunity to a hide few chunks of wall chippings which was like chalk. He preoccupied himself by math problems sketched on the floor. Initially, he had a *kilim* (rug), but that was taken away. He used his shoes as headrest, but that too was taken away, as was his clothing, leaving him with underclothing. The small opening above the doorway was boarded up, preventing natural light. The cell was damp, infested, and the walls moldy. All this is based on his assertions in his defense plea. Soon he began to feel the chill, and in December came the indictment, move to the group cell (*Falak'e*) and three days thereafter a transfer to *Qasr* prison.[6]

Others from the "fifty-three" had also stayed in the detention center all this time, about three months, a mix of solitary and group cell facing the courtyard, called *Falak'e* (the cell having the same name). From the time Taghi Erani was taken to cell no. 28, his contact with the rest had reduced dramatically; this was in September. Occasionally, he would be taken out, as he does make reference to presenting the mold in his cell to the staff at the clinic. Overall, however, it was a different treatment than the others received.

The authorities hesitated to transfer the group to *Qasr*, as they did not want to mix political and non-political prisoners (risk of politicization), neither did they want to mix this group with the veteran political activists, who could offer coaching and consolation. Some had been there since 1929, in the aftermath of the APOC strike; others since 1930, after the Agabekov defection. None had had a trial or fixed sentence; they were there indefinitely. This included a mix of old communists (Stalinists, Trotskyites, independents) and labor-union activists.[7] At any rate, a decision was made to make the transfer, possibly to open up room for incoming detainees.

***

The year 1938 was eventful. It began by Taghi Erani's transfer to *Qasr* prison, along with the others. He was consigned to ward two in a group cell. This is where the notable characters of the "fifty-three" were held, including the pilot and his entourage, a group from Qazvin, which is where he had begun his underground career. The Berlin-educated medical doctor was also in this cell. This ward was connected to ward seven which housed veteran political prisoners and so-called bourgeois tranche of the "fifty-three," those from well-off families, including the Grenoble-educated lawyer and inspiring novelist. And there was ward four, the three being interconnected, where the rest of the "fifty-three" were held (so-called "proletarians," the print-setter and locomotive conductor, and many others, as well as the students). This ward also held a contingent from Rasht, the so-called twenty-three, which had triggered the arrest of the "fifty-three." During certain periods, it was possible to move freely between these three wards, as the guards would leave the cell doors unlocked. These wards shared a common courtyard.

Goals pursued with respect to Taghi Erani were to crush him psychologically, and to preserve him physically, until trial was completed. Transfer to the group cell (*Falak'e*) at the detention center, and to *Qasr* prison and to this particular ward and cell had the goal of confronting him with the rest of the group who were convinced of his guilt in collaborating with the police. It was taken as a matter of fact that he had been the first to be arrested (though he was the sixth, the fifth being the pilot, the principal informer, the first the smuggler, the second his brother and so forth, the seventh the medical doctor, the eighth the pilot's right-hand man running the motel). He denied the accusation and said that there are many of you that I do not even know, how could I have turned you in (notably those from Qazvin, the inner circle of the pilot), but this was brushed off.[8]

Someone that he did know, and they had been close (a fellow who had studied chemistry in Berlin, after Erani, but returned to Persia before completing his studies, Khalil Maleki), physically assaulted Taghi Erani, and from then onwards Taghi Erani mingled mostly with those that had nothing to do with the "fifty-three." This included Alexander the Armenian with whom he set up a prank on others involving hypnosis (Taghi Erani putting Alex in a fake trance, who pretended that his sixth sense revealed to him the Ghazal that a third collaborator had picked from the *Divan* of Hafez, messaged to him by hand signals; it was a way of pulling everyone's leg as to how foolish they were to believe such nonsense, a tit for tat of sorts). He also befriended Grigorije the

Ukranian or Czech, who had rolled into Persia from the Caucasus, with whom he chatted in German.[9]

Taghi Erani was kept at *Qasr* for only sixteen days and was taken back to the detention center. It is not clear what triggered this. Either there was no particular trigger, the authorities felt their objective to destabilize him had been achieved, or possibly because he was intransigent with the guards, as a way of proving that he is not the type to collaborate with the police. Thus, he was taken back to the detention center, ward three, cell no. 28.

This time he was kept for eight months and was transferred back to *Qasr*, after the collective report-reading and assignment of defense attorneys. The main idea, perhaps, was for him to observe how docile others had become, enticing him to throw in the towel and plead for mercy at the trial. But for whatever reason, perhaps because he was convinced that he would be eliminated, he stayed intransigent. The prep session for the trial had put him in high spirits, showing that he had not collaborated with the police, even sheltered the others, in contrast to the pilot.

During this stay, he had interactions with the working-class types of the "fifty-three," ward four. One asked if it is alright to spy for the Soviets. He paused and said an abode needs both an architect and a janitor (*ken'nas*); why should we act as janitors for others while we can act as architects for ourselves. He was quizzed on how to react if the Red Army invades Persia; to which he said what do you do when someone breaks into your home. He was asked about Trotsky and Trotsyites; he said we do not really know what it is, and it is not our problem anyway. He was asked about the Moscow show trials. To which he said we do not know, perhaps they are innocent. And as a morale booster on state of affairs in Persia, he gave the cultivator and crow analogy; one spreads thousands of seeds, the other cannot unearth but a few. A bombshell put an end to this ambience, and within days he was back to the detention center, this time, for harsh punishment. Dropped was the plan to keep him there until the trial was done and over with.[10]

With the new director at *Qasr* the rules had changed; no more home food, no more heating elements, no more reading materials, and early blackout. One Friday night, the lights were put out early, but the guards kept speaking in a high voice. A prisoner in ward two (the one who had studied chemistry in Berlin, Khalil Maleki) objected, as he was suffering because of tragic family news; his two-year old son had passed away. The guards took him out, assaulted him, and said this is no place to

"play Bolshevik." He was transferred to the detention center for punishment. This caused havoc amongst the prisoners. The next day ward
two decided to launch a protest by a hunger strike, Taghi Erani leading the pack, the pilot backing him, as did most of the others. For the
pilot, it was a way to recover from the embarrassment of the confession
reading day. Ward four was consulted and many agreed to join; likewise,
ward seven, but many of the "fifty-three" in this ward stayed out, though
the veteran political prisoners who had been there for years and had lost
many perks with the new rules decided to do common cause, despite fac-
tional differences amongst them.[11]

Sunday the food trays were pushed back, and some even went dry
(so-called Chinese style), refraining from water. The news spread rapidly, sixty had joined the strike; by Monday it went to one hundred and
caused the poet–prisoner, Farokhi-Yazdi, to compose a few verses (he
was in a different ward and was not part of the strike). At any rate, nearly
half of those in wards two, four, and seven had joined the strike. News
was conveyed through the prison and police hierarchy to the court, dictates came down to crush the strike.[12]

Tuesday strikers let the families know of what was going on (visitation day), urging them to go to *Majles* and alert the deputies that prison
authorities are jeopardizing prisoners' lives. The same evening over one
hundred commandos descended on *Qasr*. Ten strikers were taken away
as the ringleaders: Taghi Erani, the pilot, the Berlin-educated medical
doctor, and seven others. They were taken to the prison director. They
were assaulted and told this is not Spain here (where Republicans were
still holding their own). Seven of the ten were taken to a courtyard
(called *Bagh*, orchard), given an intense lashing, which triggered various
mottos. They were then sent to the detention center. The seven plus the
one whose son had passed away were subject to further physical punishment. Five were returned on Wednesday, but the strike continued.[13]

On Thursday, the Grenoble-educated lawyer who was at *Qasr* in ward
seven was summoned to the prison director's office. He returned to ward
seven and urged that the strike be terminated. One of the cleaners also
said that wards two and four have already stopped. This was not true, but
the majority in ward seven stopped and eventually wards two and four
stopped as well. Saturday was the last day of strike, seven days all told.
Two more returned from the detention center on that day, the one whose
son had passed away and the pilot. One remained, Taghi Erani. All this
happened in September 1938, and six weeks later the trial was held.[14]

***

We have already given a description of the trial, held in the *haram* of Fat'h-Ali Shah, with the Qajar artwork depicting Russo-Persian wars on the ceiling, and a staged arrangement in which the sentences seem to have been cut out beforehand. Taghi Erani was called for his defense on the last day of the proceedings (November 12, 1938), before the closure, and in short, he condemned the 1931 anti-socialist law as a violation of Persia's constitution which guaranteed freedom of opinion and expression. He derided the legislative branch and the judiciary and its representatives in the court for slavish subservience to the executive branch, hijacked by the police and intelligence service, and stated that the root cause of this show trial is threefold—intelligence service's desire for aggrandizement, imperatives of domestic policy, and exigencies of international relations. He gave an elaborate description of the first two. He declared that his role is not to defend himself, but to defend the fundamental principles, and victims of violation of these principles, including the diverse population he had come to know about in the prison (women, regular prisoners, worker activists, prisoners of conscience, held indefinitely, subject to medieval treatment), including the "fifty-three," taking exception to the pilot who had produced an "encyclopedia" for the police. As for himself, and why he was made out to be the leader, he indicated that if what is attributed to him is true—that by the movement of one arm mountains move, by the other seas surge—he would be justified in claiming prophethood; but there was no such thing—he was a patriot, scholar, scientist, administrator, and careful review of his life record proved this without a shadow of a doubt. This trial, which is not unlike what has been observed recently in Hungry (Rakosi) and Germany (Reichstag fire), rekindling inquisitions of distant past (Socrates, Giordano Bruno, Galileo), is being watched by the world. This indictment, this defense, and this verdict, all three shall be etched in history, so beware. Do such as not to be cast as laggards of civilization.[15]

This was in November 1938, sentences fell on the fifteenth. He was then transferred to the detention center and that was the last that was seen of him by the "fifty-three." In fifteen months, prisoner no. 740, Taghi Erani, would perish at the detention center, graphic eyewitness accounts extant;[16] after his demise, his legacy the only thing remaining. Taghi Erani's mother passed away in 1946, having been spared the amnesty of the executioners (1947). The sisters passed away after the 1979 revolution—Kokab (1989), Irandokht (1991), and Shokat (2012).

The last trace we have of Mash'hadi-Reza Damavandi, the house-servant of the Erani family, is from August 30, 1937, a few months after the arrest of Taghi Erani, being interrogated by the police. He was married and had three children; we know no more.[17]

LEGACY

Taghi Erani's early life and outlook was shaped by the realities of Qajar Persia. The dichotomy of Qajar was that on the one hand it represented a lethargic political order resting on serfdom and tribalism, glued by a fief-based pact with local magnates (*Tiyul*) and an alliance with the *Shi'a* clergy, keeping the country in a state of chronic underdevelopment, vulnerable to external agendas. On the other hand, from within Qajar had sprouted healthy instincts for rejuvenation in the form of a constitutional movement and a program of modernization dating back to crown price Abbas-Mirza, continuing with Amir Kabir, and with progressive forces of the constitutional revolution (*Hezb'e Demokrat*), reforming the state apparatus, a mild Meiji—customs, finance, gendarmerie, education, and so forth. This was the syncretism of Qajar.

Diametrically opposite views emerged on the ensuing course of development of Persian society, triggered by stark historical realities (Persia as battleground of WWI, devastation and depopulation)—on the one hand total dispensation with Qajar, including its achievements in the realm of civil society (constitution, free press, political parties, authentic *Majles*, and the drive by Qajar sages to relegate the monarchy to a ceremonial function, emulating a European model), and on the other a monarchic system with the king acting as a proxy for the nation, backed by a military class, financed by exactions from the population, keeping the nation in check, without any safeguards to keep the monarch in check, including rampant cronyism and dubious transactions with foreign powers (concessions, military bases, oil, port projects, railways, arms purchases).

Taghi Erani was vehemently opposed to the latter course of development, the coronation of Reza Shah propelling him to decisive political action. In the meantime, he picked up insights from his life experience in Germany, which was short but intense and penetrating. On the one hand, appropriation of know-how in fundamental science and technology as an imperative in architecting a modern nation (existential modernization, in the absence of which there remains only the façade and semblance of modernity), and on the other, the necessity to radically transform the capitalistic social model, the breakdown of which had caused the Great War and was once again leading to militarism, fascism, and belligerency menacing to human existence. His personal observations in Europe in the 1930s, rearmament and girding for war, fueled this assessment. Thus, he subscribed to notions of radical pacifism, deep social rehabilitation, and state-led modernization.

Upon return to Persia, he was able to create a cultural movement conductive to this line of thinking, as epitomized by his epoch-making journal (*Donya, Le Monde*) and ambitious scientific and cultural publication projects, but he was not able to transform this into a social movement to radically change the political landscape. His reliance on the intelligentsia, the fickleness of which he alluded to in his defense plea, the inability to connect with the underprivileged sectors of society, those whom he considered to be the engines of change, and operating as an insider of the Pahlavi order but with an outsider agenda meant that the slightest error would be fatal. In this setting, he committed tactical errors, precipitating his demise.

In terms of Taghi Erani's intellectual legacy, three aspects may be highlighted. First is related to the concept of volition, central to Wundtian psychology, of which he was a disciple. It maintains that human action is a vital component of evolution of human society (physical environment, geography, the other component), and it is motivated by biological, social, and existential instincts. It is the idea that posterity is not preordained and is shaped by human intervention. This he demonstrated by his alertness to developments he considered pathological in Persia's history, notably the emergence of Pahlavi out of the ruins of Qajar, which led to a system that was modern in appearance but medieval in substance: a view he forged from the inside. Likewise, it made a radical pacifist out of him, as he was situated in a time horizon where not only could he observe the ruins of the Great War but could also see the unfolding of the next (degeneration of Weimar to the Third Reich); active pacifism at the national level the only recipe to abort this process. Progressives in each country having the existential-humanist responsibility of forestalling their home governments that have a propensity to wage war (e.g., Karl Liebknecht, defying the Kaiser).

Second is related to the notion of what renders human action intelligent, that is, beneficial to the prospects of human existence. Here, he lays stress on the primordial nature of logic, which flows from human endeavor and experience in nature, society, and the self (introspection).

Logic is a distillation of the insights derived from these endeavors, latent in natural and human sciences, careful study of which is vital to honing one's logical faculties, which reveal interactions in space, evolution in time, transformation in matter, destroying all fixed categories and dichotomies, including our concepts of such fixities. This he calls the logic of dialectics, which is universal, operative in the inanimate and animate worlds, which are interwoven, as it is in nonhuman and human kingdoms. For him, the more we accustom ourselves to think in terms of flux—emergence of new from old, preservation of past in the present, change as genealogy, tension between immediate and evolutionary causation, long and short cycles, spatial interactions, near field and far field, flow of matter, including human hordes, their artifacts and mental products, spanning the vast manifold of society–geography–history, as a reflection of matter-space-time—the more vividly we can capture the workings of natural and human phenomena. Application of this logic is more important than adherence to particular set of precepts or ideology ("isms").

Thus, what matters most is not the quantity of data to which we are exposed but the quality with which we process this data and glean from it its essential characteristics using our logical acumen. This is valid for individuals as it is for communities and societies, latter in the form of mental culture. What diverse groups achieve in a given environment is a function of the quality of their mental culture, more than characteristics of the physical environment, for we observe widely varying results obtained from similar environments by different cultures.

This calls for constant refining of mental culture, purging its decayed tissue, language, and preservation of its lucidity a critical element to preserve the force and vitality of a culture. These concepts are very present in his works. We may add to this the observation that the statesmen of the past had peculiar ways of putting this principle of cultural rejuvenation to work—deliberate migration policies and bringing in new cultures into the existing mix (e.g., infusion of Armenian entrepreneurial know-how from Caucasus into central Persia by Shah Abbas, boosting the silk trade via the Persian Gulf, settlement of peoples of Chinese Turkestan, Kashgar region, also in central Persia, leading to flourishing of Safavid porcelain industry), or Russian monarchs and infusion of German settlers to push the boundaries of an expanding empire. In short, versatile logic for the individual, vibrant mental culture for the community is indispensable to intelligent human action, hence, the primordial role of education and pedagogy, a prime area of application of insights from psychology.

The third is related to how this living logic finds its expression in different domains of human action. He has much commentary on how modern physics, chemistry, and biology—the so-called exact sciences—are nothing but the manifestation of this universal logic in the realm of nature, simply because this logic originates from these domains and disciplines in the first place.

Likewise, this logic has its manifestation in the realm of human society, notably social and political philosophy. Here, his accent is on the notion of nation-state, which is a product of historical evolution, thus not an immutable entity, not existing since times immemorial, yet demonstrating resilience and longevity, however not without existential crises.

His take on dealing with this crisis is not to discard the concept of nation-state but to reengineer it: hence, his narrative that modern civilization must be national in scope, but international in spirit. And the pillars of this nationalism with an internationalist mind-set rest on the trinity of appreciation, creation, and distribution of wealth, wealth as all that satisfies human wants and necessities.

The first element is cultural awareness of the primordial nature of science and technology to wrest from nature man's basic necessities. An appreciation that without this fixation with technical savviness, mastery, and innovation, we sink to ever lower rungs of civilization. Modern crises, according to Taghi Erani, are not caused by an excess of technical innovation but by its irrational and erratic application. There is a clear echo of Wundt here.

Contained in the concept of wealth creation is the concept of exchange and trade (CRRP accent on mercantilism), for the human material and the physical environment have a natural diversity across the globe, necessitating exchange relationships. And contained in the concept of trade and exchange, notably in modern times, we may add, is the concept of *breathing space*, to have access to movement and connectivity across lands and seas. Leading nations are distinguished by their monopolization of highways of global exchange, leading to inordinate appropriation of wealth to the detriment of those deprived of such access.

What explains the rise and fall of different regions of the globe across the historical timescale, according to this inference, is largely the capitalization of the emerging pathways that render older ones insignificant (silk road superseding ancient trade routes, marine routes connecting Europe to Asia, superseding the silk road, and in recent times the northern sea routes becoming a viable alternative to the southern routes, all these underpinning the civilizational seesaw across the Eurasian landmass from antiquity to the present).

Resurgence of nationalism in trailing nation-states manifests itself as a bid to gain a fair share of this global pathway as there is little point in pursuing vigorous wealth production within confines of one's country, unless conduits for regional and international trade are secured. And this highlights the nexus of merchant shipping and naval strategy (China and its marine strategy, Iran and access to east Mediterranean through allies and to Indian Ocean through port projects).

And finally, distribution of wealth, enshrined in covenants, pacts, deals, social contracts, in the bygone era and in legislation and fiscal and monetary policy in modern times, subscribing to hard-earned societal values, solidarity and humanism, is the bedrock of social stability, yet it is the resultant of an ongoing tug of war between special interest groups, classes, factions, castes, a strife reaching equilibrium only to be thrown into disequilibrium before reaching another equilibrium (Rudolf von Jhering's notion of jurisprudence as a reflection of this strife). This tug of war within society is what determines its internal solidity, hence vulnerability to external threat and predatory forces. Dialectics of spatial interaction produces the phenomenon of suppression of social volition for change in regions in turmoil. Challenges to an unjust social contract remain subdued when the region in which the nation is embedded is in the throes of chaos. Such is the outlook today in the Middle East.

The corollary to this principle of distribution of wealth as bedrock of social stability is that if it is abstracted from the principle of wealth creation, it degenerates to the principle of distribution of poverty, a downward spiral, just as when abstracted from the principle of appreciation of wealth, reduces to systemic stagnation, primitivism. Thus, these principles do not have independent validity; they are valid only in combination.

Therefore, in terms of political philosophy, much can be deduced from the concept of the trinity of wealth appreciation, creation, and distribution, when applied across historical timescale and global geography, within the realm of nation-state, allowing the identification of fields of action ripe for intelligent intervention, to steer the course of events toward social progress. This process has neither a preordained path nor a predetermined outcome, making intelligent human action the vital element, at individual and collective scales, notably combating poverty and propensity for wars of domination and plunder. Such is the legacy of Taghi Erani in the sphere of political philosophy, deduced from his notions of centrality of volition and dialectical logic, which according to him, trumps all ideologies. In short, his was a brief life pregnant with fertile ideas, an infinity hidden in a finite magnitude; in the words of Rumi—*through the eye of the needle, a turning night of stars.*

NOTES

1. Descriptions of *Qasr* prison and the detention center (*Zendan'e Movaghat* in *Shahr'bani*) compiled from accounts of Bozorg Alavi, Anvar Khame'i, Jafar Pishevari, Khalil Maleki, and Nosratollah Jahan-Shahlu-Afshar.
2. See Momeni, *Parvandeh*, pp. 14, 17, 28–29, 35, 38, 225–263, and Boroujerdi, *Erani Faratar az Marx*, p. 322, for Erani interrogations.
3. See Momeni, *Parvandeh*, pp. 235–261, for Erani confessions after the police shared Kam- Bakhsh's revelations.
4. See Momeni, *Parvandeh*, pp. 294–297, for police requisitions at Erani's home.
5. See Bozorg Alavi, *Panjah'o Se Nafar*, pp. 71–74, and Khame'i, *Panjah Nafar va Se Nafar*, pp. 22, 136, 145–146, and Jahan-Shahlu- Afshar's account in Boroujerdi, *Erani Faratar az Marx*, p. 190. These are accounts of Erani's first solitary confinement.
6. See Erani testimony about second solitary confinement in Boroujerdi, *Erani Faratar az Marx*, pp. 322–323. See pp. 286–287 for testimony of Iraj Eskandari, bribed a guard to visit him there.
7. Political prisoners of the left rounded up during 1928–1931, named in period documents included—Yusef Eftekhari (trade-union leader), Rahim Hamdad, Ali Omid (alias "Gandhi"), Ardeshir Avanesian, Jafar Pishevari, Reza Rousta, Rashid Ardalan, Mir-Ayoub Shakiba, Abdol-Ghadir Azad, Dadash-Taghizadeh, Alizadeh, and Ataollah Ghadir.
8. Erani's first stay at *Qasr*, 8–24 Dey 1316; for account of group aggression toward him, see Boroujerdi, *Erani Faratar az Marx*, pp. 190–193, and Katouzian, *Khaterat'e Khalil Maleki*, pp. 280–284.
9. See Katouzian, *Khaterat'e Khalil Maleki*, pp. 283–284, for an account of his clash with Erani; see Khame'i, *Panjah Nafar va Se Najar*, pp. 143–144, 209, 216, for account of how Erani took distance from the group.
10. Erani's second stay at *Qasr*, 5–29 Shahrivar 1317; see Khame'i, *Panjah Nafar va Se Nafar*, pp. 24, 33, Bozorg Alavi, *Panjah'o Se Nafar*, pp. 137–138, and Boroujerdi, *Erani Faratar az Marx*, p. 403.
11. Hunger strike was during 27–31 Shahrivar 1317. For its spread, see Khame'i, *Panjah Nafar va Se Nafar*, pp. 154–155, and Bozorg Alavi, *Panjah'o Se Nafar*, pp. 141–147.

12. Ibid., on the order to crush the hunger strike.

13. Police assault on hunger strike, see Khame'i, *Panjah Nafar va Se Nafar*, pp. 156–157, and Bozorg Alavi, *Panjah'o Se Nafar*, pp. 148–150.

14. End of hunger strike, see Khame'i, *Panjah Nafar va Se Nafar*, pp. 158–159, Bozorg Alavi, *Panjah'o Se Nafar*, pp. 148–155, and Katouzian, *Khaterat'e Khalil Maleki*, pp. 266–275.

15. See Erani, *Defaiyat*.

16. Eyewitness account of Erani death, see Boroujerdi, *Erani Faratar az Marx*, pp. 691–698.

17. See Boroujerdi, *Erani Faratar az Marx*, pp. 459–460, police interrogation of house-servant.

BIBLIOGRAPHY

ARCHIVES

Bankverbindung
Documents of Brandenburg-Preußischen Staates from the beginning until 1945 were kept at the Bankverbindung (Staatsbank der DDR, Konto-Nr. 6666-10-180104). Documents on Taghi Erani are held under the rubric – "Tätigkeit des außerplanmäßigen Dozenten am Orientalischen Seminar in Berlin Dr. Taghi Erani, 1925–1929". Ref: Neue Akteubezeichnung Wissenseb pp. Reziehung zum Ausland, teil XVI nr. 6 b Bd. 9.

British Foreign Office
FO 371/21890—Alleged Communist Conspiracy in Persia. Mr. Seymour to Lord Halifax (3-November-1938, 23-November-1938), Reader Bullard to Anthony Eden (6-April-1944).

Bundesarchiv
———, Carl Heinrich Becker, http://www.bundesarchiv.de/aktenreichskanzlei/1919-1933/00a/adr/adrag/kap1_2/para2_72.html.
———, Bernhard Dernburg, http://www.bundesarchiv.de/aktenreichskanzlei/1919-1933/00a/adr/adrag/kap1_4/para2_62.html.
———, Wilhelm Litten, http://www.bundesarchiv.de/aktenreichskanzlei/1919-1933/00a/adr/adrhl/kap1_5/para2_153.html.
———, Friedrich Rosen, http://www.bundesarchiv.de/aktenreichskanzlei/1919-1933/00a/adr/adrmr/kap1_6/para2_163.html.
———, Wilhelm Solf, http://www.bundesarchiv.de/aktenreichskanzlei/1919-1933/00a/adr/adrsz/kap1_1/para2_385.html.

California Death Records Index, 1940–1997, http://vitals.rootsweb.ancestry. com/ca/death/search.cgi.

Deutschen-Nationalbibliothek

———, Rosen, https://PORTAL.DNB.DE/OPAC.HTM?METHOD= SIMPLESEARCH&QUERY=118749730.

———, Schabinger-von-Schowingen, https://portal.dnb.de/opac.htm?method= simpleSearch&cqlMode=true&query=idn%3D106835483.

Erani Family Archives

Includes family genealogy, photographs, and Taghi Erani's invitation cards to the Golistan palace, and invitation card from Wipert von Blücher for debut of a Third Reich propaganda movie.

Friedrich-Wilhelms-Universität zu Berlin

Erani, Tagi, Die reduzierenden Wirkungen der unterphosphorigen Säure auf organische Verbindungen, Inaugural-Dissertation zur Erlangung der Doktorwürde genehmigt von der Philosophischen Fakultät der Friedrich-Wilhelms-Universität zu Berlin, Von Tagi Erani aus Teheran (Persien), Tag der Promotion: 19-Dezember-1928; Druck: Sonnendruckerei G.m.b.H., Charlottenburg 4, 1928.

Verzeichnis-Personal (31-Dezember-1925 zu 31-Dezember-1928); 1925– Amtliches Personalverzeichnis der Friedrich-Wilhelms-Universität zu Berlin, Abgeschlossen: 31. Dezember 1925; p. 42, under III. Außerplanmäßige, nebenamtliche Dozenten: "Erani, Taghi, beauftragt mit einer Vorlesung über orientalische Rhetorik, Stilistik und Logik – Bl-Charl., Weimarer Str. 27." 1928–Amtliches Personalverzeichnis der Friedrich Wilhelms-Universität zu Berlin für das 119. Rektoratsjahr, Rektor: His, Abgeschlossen: 31.Dezember 1928; p. 49: "Erani, Dr. Phil. (Orientalische Rhetorik, Stilistik und Logik) – W 15, Fasanenstr. 22".

Verzeichnis-Vorlesungen (Wintersemester 1925/1926 zu Wintersemester 1928/1929); Friedrich-Wilhelms-Universität zu Berlin, Verzeichnis der Vorlesungen; Wintersemester 1925–26, vom 16-Oktober-1925 bis 15-März-1926, Orientalische Rhetorik, Stilistik und Logik, Taghi Erani, Mo Do 7–8 abends, p. 83. Wintersemester 1928–29 vom 16-Oktober-1928 bis 15-März-1929, Orientalische Rhetorik, Stilistik und Logik: T. Erani, Mo Do 19–20, p. 106.

Gedenktafel-Tagi-Erani

Tafel am Haus Fasanenstraße 22, 10719 Berlin: "In diesem Hause lebte und arbeitete in den zwanziger Jahren der iranische Wissenschaftler und Antifaschist; Dr. Tagi Erani; 1903–1940." https://www.berlin.de/ba-char-lottenburg-wilmersdorf/ueber-den-bezirk/geschichte/gedenktafeln/artikel.125704.php.

Nauka, Department of Oriental Literature

'Zhizni Dannyy Svoboda' (Lives Dedicated to Freedom), Moscow, 1964.

<u>Sydney Morning Herald</u>
'Battle at Tabriz, Bombardment and Sortie, Nationalists Rout Royalists, Women as Soldiers', 22-December-1908, Trove, http://trove.nla.gov.au/ndp/del/article/14984090.

PRIVATE CORRESPONDENCE

Alavi, Bozorg, seven letters addressed from 1035 Berlin DDR, Franfurter Allee 2, spanning January–November 1986.
Ayromlou, forename not specified, letter dated 1-Farvardin-1366 (21-March-1987) from Katterbachstrasse 108, 5060 Bergisch Gladbach 2, W. Germany.
Azar, Mehdi, first letter dated 13-Ordibehesht-1366 (3-May-1987), from Harris Tower, No. 4135, 1-1 Roppongi 2 Chome, Minato-Ku, Tokyo 106, Japan; letter two dated 4-June-1987, from 421 Ridgeley Road, Norfolk, VA 23505, USA.
Jamalzadeh, Seyyed M.A., two letters from 78 Florissant, 1206 Genève, dated 23-May-1986 and 4-Azar-1365 (25-November-1986).
Khame'i, Anvar, letter dated 5-Farvardin-1365 (25-March-1986) from Tehran, P.O. Box 13145-339, in letterhead from his former academic position (Dr. Anvar Khamei, Professeur de Statistiques et d'Economie; Université Lovanium, P.B. 177, Kinshasa 11, Congo).
Shokat, Parviz, 28-January-2011 correspondence, he met with Dr. Shokat Erani in 2008 and examined some of Taghi Erani's manuscripts, in one with his scribbles about his journey to Tajrish.
Wertheimer, Michael, correspondence about Wolfgang Köhler, 17-August-2016.

THESES

Mohammadi-Doostdar, Alireza, *Fantasies of Reason: Science, Superstition, and the Supernatural in Iran*, Doctoral Dissertation, Harvard University, Cambridge, 2012.
Petersson, Fredrik, *'We Are Neither Visionaries nor Utopian Dreamers'—Willi Münzenberg, the League Against Imperialism, and the Comintern, 1925–1933*, Doctoral Thesis, Åbo Akademi University, Turku, Finland, 2013.

PUBLISHED SOURCES IN ENGLISH, GERMAN, FRENCH

Abrahamian, Ervand, *Iran Between Two Revolutions*, Princeton, 1982.
———, 'Taqi Arani', *Encyclopaedia Iranica*, 1986.
Afshar, Iraj, 'Kāva Newspaper', *Encyclopaedia Iranica*, 2013.
Agabekov, Georges, *O.G.P.U., the Russian Secret Terror*, New York, 1931.

Amanat, M., 'Mohammad-Hosayn Ayrom', *Encyclopaedia Iranica*, 1987.

Arberry, A.J., *Razi's Traditional Psychology*, Islamic Book Service, Damascus, n.d.

Asmussen, Jes P., 'Christensen, Arthur Emanuel', *Encylopaedia Iranica*, 1991.

Azizi, Mohammad Hossein, 'A Brief History of Smallpox Eradication in Iran', Academy of Medical Sciences of the I. R. of Iran, Tehran, 2009.

Bajanov, Boris, *Bajanov révèle Staline*, Gallimard, Paris, 1977.

Banani, Amin, *The Modernization of Iran, 1921–1941*, Stanford, 1961.

Bast, Oliver, 'German-Persian Diplomatic Relations', *Encyclopaedia Iranica*, 2001.

Bosworth, C.E., 'Arran, a Region of Eastern Transcaucasia', *Encyclopaedia Iranica*, 1986.

Botting, Tom, 'Ode to Amir Abu Nasr Mamlan, Qasideh by Qatran al-Tabrizi', in The Cambridge History of Iran, Vol. 4, R. N. Frye (ed.), 1975.

Breitscheid, Rudolf, *Antifaschistische Beiträge, 1933–1939*, Marxistische Blätter, Frankfurt, 1977.

Bridgeman, Reginald, 'Persia and British Labour, July 1931', in The Left in Iran, 1905–1940, Cosroe Chaqueri (ed.), Merlin, Pontypool, 2010.

Brook-Shepherd, Gordon, *The Storm Petrels, the Flight of the First Soviet Defectors*, Ballantine, New York, 1977.

Browne, E.G., *The Persian Revolution of 1905–1909*, Cambridge, 1910.

———, 'Some Recent Persian Books Publications of the Kaviani Press', *Journal of the Royal Asiatic Society of Great Britain & Ireland* (New Series), Vol. 56, No. 2, pp. 279–280, April 1924.

———, *A Literary History of Persia*, Cambridge, 1906 (Vol. 1) to 1924 (Vol. 4).

Calafi, Farnaz, et al., 'Iran's Yankee Hero', *New York Times*, Op-ed, 18-April-2009.

Chaqueri, Cosroe, 'Iraj Eskandari', *Encyclopaedia Iranica*, 1998.

———, 'Solayman Mirza Eskandari', *Encyclopaedia Iranica*, 1998.

———, *Un prince iranien rouge en France, vie et oeuvres du Iradj Eskandari*, Antidote, 2002.

———, *The Left in Iran, 1905–1940*, Merlin, Pontypool, 2010.

Coffey, Patrick, *Cathedrals of Science*, Oxford, 2008.

Cronin, Stephanie, 'An Experiment in Revolutionary Nationalism: The Rebellion of Colonel Muhammad Taqi Khan Pasyan in Mashhad, April–October 1921', *Middle Eastern Studies*, Vol. 33, No. 4, pp. 693–750, 1997.

———, *The Making of Modern Iran—State and Society Under Riza Shah, 1921–1941*, editor, Routledge, London, 2003.

———, *Reformers and Revolutionaries in Modern Iran*, editor, Routledge, London, 2004.

———, 'Popular Politics, the New State and the Birth of the Iranian Working Class: The 1929 Abadan Oil Refinery Strike', *Middle Eastern Studies*, Vol. 46, No. 5, pp. 699–732, September 2010.

————, *Iranian-Russian Encounters, Empires and Revolutions Since 1800*, Routledge, London, 2013.

de Planhol, Xavier, 'Tehran, Persian City at the Foot of the Alborz', *Encyclopaedia Iranica*, 2004.

Dzhirkvelov, Ilya, *Secret Servant, My Life with the KGB and the Soviet Elite*, Harper Collins, 1988.

Elwell-Sutton, George, *Modern Iran*, Routledge, London, 1941.

Fiore, Massimilano, *Anglo-Italian Relations in the Middle East, 1922–1940*, Routledge, Abingdon, 2016.

Fischer, Emil, *Chemical Research and Its Bearings on National Welfare*, Berlin, 1911.

Freud, Sigmund, *On Dreams*, W. W. Norton, New York, 1980.

————, *Civilization and Its Discontents*, Martino, Connecticut, 2010.

————, *The Future of an Illusion*, Collected Works, pp. 259–294, Pacific, 2010.

————, *The Ego and the Id*, Collected Works, pp. 223–258, Pacific, 2010.

————, *Beyond the Pleasure Principle*, Collected Works, pp. 183–221, Pacific, 2010.

————, *The Psychopathology of Everyday Life*, Collected Works, pp. 7–114, Pacific, 2010.

Freund, Gisèle, 'International Congress for the Defense of Culture, Hall of the Mutualité, Paris, 21st of June, 1935', Collection of Marita Ruiter, Galerie Clairefontaine, 1985.

Friedrich, Otto, *Before the Deluge, a Portrait of Berlin in the 1920s*, Fromm, New York, 1986.

Gay, Peter, *Weimar Culture, the Outsider as Insider*, W. W. Norton, New York, 2001.

Ghani, Cyrus, *Iran and the Rise of Reza Shah: From Qajar Collapse to Pahlavi Power*, Tauris, London, 2000.

Gibbon, Edward, *The History of the Decline and Fall of the Roman Empire*, Vol. 5, 1788.

Gurney, John and Nabavi, Negin, 'Dar Al-Fonun', *Encyclopaedia Iranica*, 1993.

Haffner, Sebastian, *Failure of a Revolution, Germany 1918–1919*, Banner, Chicago, 1986.

————, *The Ailing Empire, Germany from Bismarck to Hitler*, Fromm, New York, 1989.

————, *The Rise and Fall of Prussia*, Phoenix Giant, 1998.

Hallas, Duncan, *The Comintern*, Haymarket, Chicago, 1985.

Haque, Amber, 'Psychology from Islamic Perspective: Contributions of Early Muslim Scholars and Challenges to Contemporary Muslim Psychologists', *Journal of Religion and Health*, Vol. 43, No. 4, Winter 2004.

Harvard Fellows College, *Iran from Religious Dispute to Revolution*, Wisconsin University Press, Madison, 1980.

Hauptmann, Gerhart, *Phantom*, B. W. Huebsch, New York, 1922.

Hawkins, Jeff, *On Intelligence*, Owl, New York, 2004.

Hedayat, Sadegh, *The Blind Owl*, Grove, New York, 2010.

Heyns, K., 'Chemistry at University Level in West Germany', www.pac.iupac.org, 1969.

Hubel, David H., 'The Brain', *Scientific American*, W. H. Freeman, San Francisco, 1979.

Humboldt-Universität, 'A Colonial Cultural and Language Institute', https://www.iaaw.hu-berlin.de/en/southasia/profile/history.

Irandokht, *True Translation of Hakim Omar Khayyam's Robaiyat*, Zamaneh, San Jose, 1994.

Ironside, Edmund, *High Road to Command: The Diaries of Major General Sir Edmund Ironside, 1920–1922*, Lord Ironside (ed.), Leo Cooper, London, 1972.

Issawi, Charles, *The Economic History of Iran, 1800–1914, with an Addendum on Iran's Economy During 1914–1970*, University of Chicago, 1971.

Ivry, Alfred, 'Arabic and Islamic Psychology and Philosophy of Mind', *Stanford Encyclopedia of Philosophy*, 2008.

James, William, *Talks to Teachers on Psychology and to Students on Some of Life's Ideals*, Dover, New York, 2001.

Javadi, Hassan, 'Mahmud Ganizada', *Encyclopaedia Iranica*, 2000.

Jenkins, Jennifer, 'Hjalmar Schacht, Reza Shah, and Germany's Presence in Iran', *Iran Nameh*, Vol. 30, No. 1, pp. XX–XLVI, Spring 2015.

Johnson, Jeffrey, *The Kaiser's Chemists, Science and Modernization in Imperial Germany*, University of North Carolina, Chapel Hill, 1990.

Jolland, Michel, 'La mort de Willi Münzenberg, Zones d'ombre et Questionnements', *le bulletin mensuel de l'Académie Delphinale*, pp. 49–66, février 2014.

Kapuscinski, Ryszard, *Le Shah*, Flammarion, Paris, 2010.

Khatib-Shahidi, Rashid, *German Foreign Policy Towards Iran Before World War Two*, Tauris, New York, 2012.

Koch, Stephen, *Double Lives, Stalin, Willi Münzenberg and the Seduction of the Intellectuals*, Harper Collins, London, 1996.

Kremers, Edward, 'Hermann Thoms', *Journal of the American Pharmaceutical Association*, Vol. XII, No. 7, July 1923.

Ladjevardi, Habib, *Labor Unions and Autocracy in Iran*, Syracuse University Press, 1985.

Lange, Christian and Mecit, Songül (eds.), *The Seljuqs—Politics, Society and Culture*, Edinburgh University Press, 2012.

Leaman, Oliver, *A Brief Introduction to Islamic Philosophy*, Polity, Cambridge, 1999.

Levenson, Thomas, *Einstein in Berlin*, Bantam, New York, 2004.

Long, Doug, 'Einstein and the Atomic Bomb', www.doug-long.com.
Lorenz, Konrad, *Behind the Mirror, a Search for a Natural History of Human Knowledge*, Harvest, New York, 1977.
Luoma, Todish, 'Review of True Translation of Hakim Omar Khayyam's Robaiyat', http://www.amazon.com/True-Translation-Hakim-Khayyams-Robaiyat/dp/1882029747?ie=UTF8&*Version*=1&*entries*=0&show-DetailProductDesc=1#iframe-wrapper.
Machiavelli, Niccolo, *The Prince*, Everyman, London, 1995.
Mahrad, Ahmad, *Die deutsch-persischen Beziehungen von 1918–1933*, Lang., Frankfurt, 1974.
Majd, Mohammad Gholi, *Great Britain and Reza Shah—The Plunder of Iran, 1921–1941*, University Press of Florida, 2001.
Mandler, George, *A History of Modern Experimental Psychology, from James and Wundt to Cognitive Science*, MIT, Cambridge, 2007.
Marchand, Suzanne L., *German Orientalism in the Age of the Empire, Religion, Race, and Scholarship*, Cambridge, 2010.
McBryde, Michael, 'Wilhelm Solf, Governor of German Samoa', http://www.goethe.de/ins/nz/en/wel/kul/mag/wwi/sam/12799761.html.
McMeekin, Sean, *The Red Millionaire: A Political Biography of Willi Münzenberg, Moscow's Secret Propaganda Tsar in the West*, Yale, New Haven, 2003.
Metzger, Rainer and Brandstätter, Christian, *Berlin, the Twenties*, Abrams, New York, 2007.
Mirabedini, Hasan, 'Bozorg Alavi', *Encyclopaedia Iranica*, 2009.
Morgan-Shuster, William, *The Strangling of Persia*, 1912.
Nasr, Seyyed Hossein and Leaman, Oliver (eds.), *History of Islamic Philosophy*, Routledge, London, 1996.
Neukrantz, Klaus, *Barricades in Berlin*, 1933, Banner, Chicago, 1979.
Nobel, *Nobel Lectures, Chemistry, 1901–1921*, Elsevier, Amsterdam, 1966.
Nye, Mary Jo, *Before Big Science, the Pursuit of Modern Chemistry and Physics, 1800–1940*, Harvard, Cambridge, 1996.
Orwell, George, *Homage to Catalonia*, Harvill Secker, London, 1938.
Planck, Max, *Scientific Autobiography and Other Essays*, Philosophical Library, New York, 1949.
———, *Where Is Science Going*, Ox Bow, Connecticut, 1981.
Rathenau, Walther, *The New Society*, Williams and Norgate, London, 1921.
Reiss, Tom, *The Orientalist— In Search of a Man Caught Between East and West*, Vintage, London, 2006.
Reza, Enayatollah, 'Arran, the Real Name of Azerbaijan', posted 28-February-2010, http://www.iranchamber.com/geography/articles/arran_real_azerbaijan.php.
Rieber, R.W. (ed.), *Wilhelm Wundt and the Making of a Scientific Psychology*, Plenum, New York, 1980.

Robinson, Daniel N., *An Intellectual History of Psychology*, University of Wisconsin Press, Madison, 1995.

Rosen, Frederic, *The Algebra of Mohammed Ben Musa Khuwarizmi*, Oriental Translation Fund, London, 1831.

Rosen, Friedrich, *Modern Persian Colloquial Grammar*, Luzac, London, 1898.

————, *Die Sinnsprüche Omars des Zeltmachers*, Stuttgart, 1922.

————, *The Quatrains of Omar Khayyam*, E. P. Dutton, New York, 1929.

————, *Oriental Memories of a German Diplomatist*, E. P. Dutton, New York, 1930.

Roth, Joseph, *The End of the World, Journalistic Writings, 1924–1939*, Hesperus, London, 2013.

————, *What I Saw, Reports from Berlin, 1920–1933*, Granta, London, 2013.

Russell, Bertrand, *A History of Western Philosophy*, Simon & Shuster, New York, 1972.

Rypka, Jan, *History of Iranian Literature*, D. Reidel, Dordrecht, 1968.

Saadi, Sheikh Muslihu'd-din, *Golistan*, Octagon, London, 1979.

————, *Bustan*, IndoEuropean, Los Angeles, 2012.

Schayegh, Cyrus, *Who Is Knowledgeable Is Strong—Science, Class, and the Formation of Modern Iranian Society, 1900–1950*, University of California Press, Berkeley, 2009.

Schmitt, Rüdiger, 'Grundrissen der Iranischen Philologie', *Encyclopaedia Iranica*, 2002.

Schütt, Hans-Werner, *Eilhard Mitscherlich, Prince of Prussian Chemistry*, American Chemical Society, 1997.

Seelig, Carl (ed.), *Albert Einstein, Ideas and Opinions*, Bonanza, New York, 1954.

Sirotkina, Irina and Smith, Roger, *Russian Federation*, in Oxford Handbook of the History of Psychology, Global Perspectives (Ch. 20), David B. Baker (ed.), Oxford, 2012.

Spengler, Oswald, *The Decline of the West*, Stellar, 2013.

Stoltzenberg, Dietrich, *Fritz Haber—Chemist, Nobel Laureate, German, Jew*, Chemical Heritage, Philadelphia, 2004.

Stone, Norman, *World War One: A Short History*, Penguin, London, 2008.

Taraporewala, Irach J.S., *The Religion of Zarathushtra*, Madras, 1926.

————, *The Divine Songs of Zarathushtra*, Hukhta Foundation, Bombay, 1993.

Vessal, K., et al., 'Development of Radiology in Iran', *Archives of Academy of Medical Sciences Iran*, Vol. 12, No. 6, pp. 611–616, 2009.

Volkov, Shulamit, *Walter Rathenau, Weimar's Fallen Statesman*, Yale, New Haven, 2012.

von Jhering, Rudolf, *The Struggle for Law*, Callaghan, Chicago, 1915.

Wallenberg, Raoul, *Letters and Dispatches, 1924–1944*, Arcade.

Werner, Christoph, 'Adam Olearius', *Encyclopaedia Iranica*, 2008.

Wertheimer, Michael, *A Brief History of Psychology*, Harcourt College, 2000.
Wilhelm II, *The Kaiser's Memoirs, Emperor of Germany, 1888–1918*, Thomas R. Ybarra (trans.), Printed in USA, undated.
Wundt, Wilhelm, *Ethics, the Facts of Moral Life*, Cosimo, 1886.
————, *Outlines of Psychology*, George Allen, 1902.
————, *Principles of Physiological Psychology*, Swan Sonnenschein, New York, 1910.
————, *An Introduction to Psychology*, George Allen, London, 1912.
————, *Elements of Folk Psychology, Outline of a Psychological History of the Development of Mankind*, George Allen & Unwin, London, 1915.
Yarshater, Ehsan, 'Persia or Iran', *Iranian Studies*, Vol. XXII, No. 1, 1989.
Zakani, Obeyd, *Mouse and Cat*, New Humanity, Victoria, 2012.

PUBLISHED SOURCES IN PERSIAN

Abde, Jalal, *Khaterat'e Dr. Jalal Abde, Chehel Sal dar Sahn'e Ghaza'i, Siyasi, Diplomasi, Iran va Jahan*, Rasa, Tehran, 1378.
Afshar, Iraj, *Zendegi'e Toufani, Khaterat'e Seyyed Hassan Taghizadeh*, Elmi, Tehran, 1369.
Afshar, Iraj and Dehbashi, Ali, *Khaterat'e Seyyed Mohammad-Ali Jamalzadeh*, Shahab, Tehran, 1378.
Agheli, Bagher, *Teymurtash dar Sahne'ye Siasat'e Iran*, Javeedan, Tehran, 1371.
Ahmadi, Hamid, *Tarikhche Ferq'e Jomhooriye Enghelabi'e Iran va Gorooh'e Erani*, Akhtaran, Tehran, 1379.
————, *Khaterat'e Najmi Alavi*, Akhtaran, Tehran, 1383.
————, *Setareh Sorkh, Majeleh Mahiya'ne Organ'e Komit'e Markazi'e Ferq'e Komonist'e Iran (1308–1310)*, Baran, Spånga, 1992.
————, *Khaterat'e Bozorg'e Alavi*, Baran, Spånga, 1997.
Alamuti, Ziaoldin, *Fosoo'li az Tarikh'e Mobarezat'e Siyasi va Ejtema'i Iran*, Khash, Tehran, 1370.
Alavi, Bozorg, *Varagh Pareha'ye Zendan*, Tehran, 1942.
————, *Panjah'o Se Nafar*, Tehran, 1944.
Alavi, Najmi, *Sargozasht'e Morteza Alavi*, Mard'e Emrooz, Tehran, 1370.
Aryanpour, Yahya, *Az Saba ta Nima, Tarikh'e 150 Sal Adab'e Farsi*, Zavvar, Tehran, 1387.
Avanesian, Ardeshir, *Khaterate Ardeshir'e Avanesian*, Negar'e, Tehran, 1376.
Azadegan, Hemad, *Defaiyat'e Ahmad Kasravi az Sarpas Mokhtari va Pezeshk Ahmadi*, Parcham, Tehran, 1943.
Bayat, Kaveh and Tafreshi, Majid, *Khaterat'e Yusef Eftekhari*, Tehran, 1370.
Behnam, Jamshid, *Berlini'ha, Andish'mandan'e Irani dar Berlin, 1915–1930*, Farzan, Tehran, 1379.
Behrooz, Maziyar, *Shooreshiyan'e Arman'khah*, Ghoghnoos, Tehran, 1380.

Boroujerdi, Hussein, *Erani Faratar az Marx*, Tazeh, Tehran, 1382.

Chaqueri, Cosroe, *Taghi Erani dar Ayeneye Tarikh*, Akhtaran, Tehran, 1387.

CRRP—Circle of Revolutionary Republic of Iran, Documents Principally in Persian, Partly in German and French, Reproduced in Boroujerdi (Erani Faratar az Marx), Chaqueri (Taghi Erani dar Ayeneye Tarikh), and Ahmadi (Tarikhche Ferq'e Jomhooriye Enghelabi'e Iran):

———, CRRP Political Program and Letter to Comintern, Berlin, 29-March-1926.

———, Iran sous le sign de la révolution, Berlin, May 1926.

———, Letter to Georg Ledebour of USPD, Berlin, 18-June-1926.

———, Political Struggle to Establish People's Sovereignty, Berlin, 1-November-1926.

———, Manifesto (Bayan'e Hagh, Proclamation of Truth), Berlin, 1927.

———, Resolution Submitted to Conference of League Against Imperialism, Brussels, 11-February-1927.

———, Resolution Submitted to Commission of League Against Imperialism, Berlin, December-1927.

———, Statement of Anti-colonial Stance, Berlin, 19-January-1928.

———, Beyragh'e Enghelab, July-1928

———, Tract Condemning Teymurtash on Visit to Europe, Berlin, 19-September-1928.

Dehbashi, Ali, *Khaterat'e Siyasi'e Iraj Eskandari*, Elmi, Tehran, 1368.

Erani, Taghi, 'Qasideh Takht'e Jamshid', Not Extant, Tehran, 1300.

———, 'Tadghighat'e Lesani', *Iränschahr*, No. 5–6, Berlin, 1302.

———, 'Soalat'e Elmi', *Iranschähr*, No. not known, Berlin, 1302.

———, 'Nam-daran'e Iran', *Iranschähr*, No. not known, Berlin, 1302.

———, 'Qasideh dar Molood'e Hazrat'e Rasul', *Azadiye Shargh*, Berlin, 1302.

———, 'Qasideh Madar'e Meehan', *Azadiye Shargh*, Berlin, 1302.

———, *Montakhab Lataef'e Molana Obeyd'e Zakani*, Kaviani, Berlin, 1303.

———, 'Azerbaijan, Yek Masaleye Hayati va Momati'e Iran', *Farangestan*, No. 5, Berlin, 1303

———, 'Teknik, Fann va Farhang', *Iranschähr*, No. 10, Berlin, 1303.

———, *Ketab'e Vajh'e Din, Hakim Nasser-Khosrow*, Kaviani, Berlin, 1304.

———, *Badaye'e Afsah'ol Motekale'min Sheikh Muslih'ol-Din Saadi Shirazi*, Kaviani, Berlin, 1304.

———, 'Rah-Ahanhaye Alman', *Sanaye'e Alman va Shargh*, Berlin, 1304.

———, 'Tarvij'e Sari'e Elm va Honar az Gharn'e Noozdahom be Baad', *Elm va Honar*, Berlin, 1306.

———, 'Oloom va Khorafat', *Elm va Honar*, Berlin, 1306.

———, *Selsele Osool'e Oloom'e Daghigh'e, Osool'e Elm'e Physik, Jeld'e Avval, Mekanik va Hararat, Physik Omoomi va Ketab'e Dasti Baraye Mohaselin'e Oloom*, Sirousse, Tehran, 1309.

————, *Osool'e Elm'e Chimi*, Sirousse, Tehran, 1310.

————, *Teorihaye Elm*, *Sirousse*, *Tehran*, 1310.

————, *Osool'e Elm'e Physik*, *Jeld'e Dovom*, *Akoustik*, *Optik*, *Magnetism va Elektricite*, Sirousse, Tehran, 1310.

————, *Osool'e Elm'e Rooh*, *Psikologi'e Omoomi*, Sirousse, Tehran, 1311.

————, 'Takamol, Tabaiyyat be Mohit, Ers', *Donya*, No. 2, Tehran, 1312.

————, 'Atom va Baud'e Chaharom', *Donya*, No. 1, Tehran, 1312.

————, 'Honar va Materializm', with B. Alavi, *Donya*, No. 1, Tehran, 1312.

————, 'Ajsam'e Radioaktiv'e Masnui', *Donya*, No. 9, Tehran, 1313.

————, 'Bashar az Nazar'e Maadi', *Donya*, No. 9, Tehran, 1313.

————, 'Farziye Nesbi', *Donya*, No. 7–9, Tehran, 1313.

————, 'Materializm'e Dialektik', *Donya*, No. 6–8, Tehran, 1313.

————, 'Tarikh-Sazi dar Honar', *Donya*, No. 7, Tehran, 1313.

————, 'Shesh Mah Majel'e Donya va Enekas'e An', *Donya*, No. 6, Tehran, 1313.

————, 'Hoghoogh va Osool'e Maadi,' with I. Eskandari, *Donya*, No. 5, Tehran, 1313.

————, 'Zendegi va Rooh ham Maadi Ast', *Donya*, No. 4 & 5, Tehran, 1313.

————, 'Khaub va Khaub Didan', with B. Alavi, *Donya*, No. 3, Tehran, 1313.

————, 'Erfan va Osool'e Maadi', *Donya*, No. 1, 3, 4, Tehran, 1312–1313.

————, *Resalat fi Sharh ma Ashkal men Mosaderat Kitab Eqlidos*, Sirousse, Tehran, 1314.

————, 'Eghtesad va Siyasat'e Eghtesadi'ye Sal'e 1934', *Donya*, No. 10–12, Tehran, 1314.

————, 'Taghir'e Zaban'e Farsi', *Donya*, No. 10–12, Tehran, 1314.

————, 'Tarikh'e Elm', with A. Khame'i, *Donya*, No. 10–12, Tehran, 1314.

————, 'Takamol'e Mojoodat'e Zende', *Donya*, No. 10–12, Tehran, 1314.

————, 'Honar dar Iran'e Jadid', with B. Alavi, *Donya*, No. 10–12, Tehran, 1314.

————, 'Bayaniye Mah'e May', in Parvadeh by Bagher Momeni, pp. 299–301, Tehran, 1315.

————, *Defaiyat*, in Parvandeh by Bagher Momeni, pp. 515–542, Tehran, 1317.

Fami-Tafreshi, Morteza S., *Polis'e Khafi'ye Iran*, *1299–1320*, Ghoghnoos, Tehran, 1367.

Foroughi, M.A. and Khoramshahi, B., *Koliyat'e Saadi*, Doostan, Tehran, 1389.

Ghassemi, Farhang, *Sandikalism dar Iran (1284–1320)*, Mossadegh Foundation, Paris, 1364.

Golbon, Mohammad and Sharifi, Yusef, *Mohakeme'ye Mohakemegaran*, Noghreh, Tehran, 1363.

Gorgani, Fazollah, 'Be Monasebat'e Salgard'e Shahadat'e Dr. Taghi Erani', *Donya*, Fourth Series, Tudeh, Tehran, Esfand 1358.

Hashemian, Ahmad, *Tahavolat'e Farhangi'e Iran dar Dore'ye Qajari'ye va Madress'e Dar-ol-Fonun*, Institute of Geography and Cartography of Sahab, Tehran, 1379.

Hashemi-Rafsanjani, Akbar, 'Rishe'ye Ghodrat'e Enghelab', *Ettelaat*, Tehran, 28-Ordibehesht-1362.

Heravi, Mehdi, *Marxism, Hezb'e Tudeh va Goroohha'ye Chap dar Iran*, Ibex, Bethesda, 2011.

Institute-of-Political-Research, Hezb'e *Tudeh* az Sheklgiri ta Foroopashi (1320–1368), Tehran, 1387.

Irandokht, 'Notgh dar Dovomin Kongreh'e Zanan'e Shargh', Tehran, 1311, in *Khaterat'e Najmi Alavi*, pp. 56–57.

———, 'Yek Shahkar dar Assar'e Talim va Tarbiyat, Helen Keller'e Amrika'i', *Donya*, No. 5, Tehran, Khordad 1313.

———, *Banoo'i dar Hojre'e Tariki, Christiana Tsai*, Villa, Tehran, 1965.

Jamalzadeh, Seyyed M.A., *Ganj'e Shayegan*, Kaveh, Berlin, 1918.

Kambakhsh, Abdol-Samad, *Nazari be Jonbesh'e Kargari va Komonisti dar Iran*, Tudeh, Leipzig, 1972.

Kasravi, Ahmad, *Azari ya Zaban'e Bastan'e Azarbayegan*, Tehran, 1312.

———, *Tarikh'e Mashroote Iran*, Tehran, 1313.

———, *Ghiyam'e Sheikh Mohammad Khiabani*, Tehran, 1321.

———, *Din va Jahan*, Tehran, 1323.

———, *Varjavand'e Bonyad*, Tehran, 1323.

Katouzian, Homa, *Khaterat'e Siyasi'e Khalil Maleki*, Europe, 1360.

Kazemzadeh, Kazem, *Asar va Ahval'e Hossein Kazem-Zadeh Iranschähr*, Eqbal, Tehran, 1363.

Khame'i, Anvar, *Panjah Nafar va Se Nafar*, Hafte, Tehran, 1362.

———, 'Harfhaye Erani dar Iran Tazegi Dasht', interview by Mohsen Azmoudeh, 23-October-2013; posted on http://www.anlam.biz/?p=13818#hHM0yX1W59S78k3r.99.

Kianfar, Jamshid, 'Mirza Mahmud Khan Ehtesham-ol-Saltaneh, Dovomin Ra'is Majles Shora'e Melli', *Peyk'e Noor*, Vol. 3, No. 5 (supplement), Winter 2006.

Mansouri, Firouz, *Motaleat'i dar Bareye Tarikh, Zaban, va Farhang'e Azarbaijan*, 2 Vols., Hazar Kerman, Tehran, 1387.

Miroshnikov, L.Y., *Iran dar Jang'e Jahani'e Avval*, Farzaneh, Tehran, 1345.

Momeni, Bagher, *Parvandeh'e Panjah'o Se Nafar*, Hossein Farzaneh (pseudonym), Negah, Tehran, 1372.

———, *Dar Khalvat'e Doost, Namehaye Bozorg Alavi be Bagher Momeni*, Nima Verlag, Essen, 1379.

———, *Donya'ye Erani*, Khojaste, Tehran, 1384.

Motazed, Khosrow, *Si Sal ba Reza Shah dar Ghazagh-Khaneh, Khaterat'e Sadegh Adibi*, Alborz, Tehran, 1385.

———, *Polis'e Siyasi*, Alborz, Tehran, 1388.

Motevali, Abdollah, *Teymurtash va Baziy'e Ghodrat*, Madres'e, Tehran, 1381.

Pasyan, Najaf-Gholi and Motazed, Khosrow, *Zendegi'e Reza Shah Pahlavi, Az Savad-Kuh ta Johannesburg*, Sal'es, Tehran, 1388.

Peymani, Nader, *Ma va Biganegan, Khaterat'e Siyasi'e Dr. Nosratollah'e Jahan-Shahlu-Afshar*, Samarghand, Tehran, 1388.

Pishevari, Jafar, *Yaddashtha'ye Zendan*, Azhir, Tehran, 1342.

——, *Tarikhche Hezb'e Edalat*, Elm, Tehran, 1359.

Raain, Ismail, *Mirza Malkom Khan*, Safi-Ali Shah, Tehran, 1353.

Rais-Niya, *Heydar Amoughli dar Gozar az Toofanha*, Donya, Tehran, 1360.

Rosen, Friedrich, *Ruba'iyat-i Umar-i-Khayyam*, Kaviani, Berlin, 1925.

Sadeghi, Zahra, *Siyasat'haye San'ati dar doran'e Reza Shah, 1304–1320*, Khojasteh, 1387.

Sadeghi, Zia, 'Mosahebe ba Dr. Mehdi Azar', Harvard University, Center for Middle Eastern Studies, Iranian Oral History Project, Norfolk, Virginia, 31-March-1983.

Shafie-Kadkani, Mohammad Reza, *Advar'e She'ere Farsi, az Mashrootiyat ta Soghoot'e Saltanat*, Sokhan, Tehran, 1383.

Shahri, Jafar, *Tehran'e Ghadim*, 5 Vols., Moeen, Tehran, 1383.

Sheikholeslami, Javad, *So'ud va Soghoot'e Teymurtash*, Tus, Tehran, 1379.

Tabari, Ehsan, *Az Didar'e Khishtan, Yad-Nameye Zendegi*, Baran, Spånga, 1360.

Tabari, Ehsan and Alavi, Bozorg, *Zendegi'nameh va Mohakeme va Matn'e Defaiye Dr. Erani*, Tudeh, Tehran, 1326.

Tabataba'i, Seyyed Mohammad Hossein, *Osool'e Falsaf'e va Ravesh'e Realizm*, annotated by Ostad Morteza Motahhari, Sadra, Tehran, 1381.

Tohidi, Sulmaz, *Matbooat'e Komonisti'e Iran dar Mohajerat, 1296–1311*, Azerbaijan Daily, 1985.

Valizadeh, Akbar, *Etehad'e Jamahir'e Shoravi va Reza Shah, Bar'rasi Ravabet'e Iran va Shoravi Miyan'e do Jang'e Jahani*, Documentation Center of Islamic Revolution, Tehran, 1385.

Yushij, Sharagim (ed.), *Namehaye Nima Yushij*, Tehran, n.d.

INDEX

Kaufmann, Max Rudolf, 79
Kaviani, 66, 76, 77, 83, 85–87, 89,
    92, 93, 119, 121
Kazemzadeh, Iranschahr, 77, 83, 143,
    145, 151
Kekulé, August, 56
Khame'i, Anvar, 14, 15, 42, 102, 151,
    153, 192, 210, 218, 221, 224,
    225, 255, 257, 276, 277
Khiabani, Mohammad, 38, 42, 151,
    169, 171, 172
Khwarazmi, Mohammad ibn Musa,
    35, 81
KMT (Kuomintang), 173
Köhler, Wolfgang, 49, 105, 130
Kornilov, Konstantin, 106, 122
Kozlov, Dmitry Timofeyevich, 228
KPD (Kommunistische Partei
    Deutschlands), 148, 149, 186,
    187, 197, 200, 214, 231, 255
Krasin, Leonid, 219
Kuchak Khan, 39, 151, 169
Kushite, 161
KUTV (Communist University of
    Toilers of the East), 196, 225

L
Lahuti, Abol-Ghassem, 165, 238, 254
Lamartine, Alphonse, 141, 151
Langmuir, Irving, 56
Lataef, 92
von Laue, Max, 64
Lavoisier, Antoine, 56
Law of Human Action, 127, 132, 134
Law of Mental Gyration, 127, 128,
    132, 134
Ledebour, Georg, 47, 51, 148
Leibniz, Gottfried, 109, 122
Leibnizstrasse, 44, 45, 76, 83, 85, 119
Lenin, Vladimir, 45, 165, 172, 178,
    179, 189, 193, 218

Lewis, Gilbert, 56, 58
von Liebig, Justus, 54, 56
Liebknecht, Karl, 82, 185, 186, 193,
    246, 272
Lindenblatt, Kurt, 158, 177, 191, 194
Litten, Wilhelm, 44, 45, 50, 79, 80,
    93–97
Litvinov, Maxime, 179, 189, 190
Lloyd George, David, 13, 15, 228,
    237
Lobachevsky, Nikolai, 68, 98
Loewi, Otto, 124
Lorenz, Konrad, 130, 137
Lucas White King, 92
Lutze, Viktor, 47
Luxemburg, Rosa, 82, 186
Luys, Jules Bernard, 106
Lyakhov, Vladimir, 24–26

M
MacDonald, Ramsay, 15, 245
Mach, Ernst, 63, 67, 127
Maleki, Khalil, 177, 258, 267, 268,
    276, 277
Malkom Khan, 72, 83
Man and Materialism, 66, 132
Mandler, George, 122
Mannich, Carl, 48, 61
Markwart, Josef, 76, 96, 97, 100
Massihi, Jean David Khan, 72
Materialism and Dialectics, 66, 132,
    211
Matin-Daftari, Ahmad, 11, 240, 255,
    257
Mauni, 160, 168, 198
Mayak theater, 188
Mayan, 161
Mazdak, 160, 168, 198
Meiji, 53, 262, 271
Meitner, Lise, 46, 49, 61, 187
Meta-psychology, 107

Piaget, Jean, 122, 123
Planck, Max, 49, 52, 58, 62–65, 67,
    88, 105
Plasticity of Mentality, 107
Plekhanov, Georgi, 218
Poetess Jahan, 92
Pour-Reza, Mahmoud, 148, 150, 152,
    153, 165
Priestley, Joseph, 56
Prince Albert, 56
Pringsheim, Peter, 49, 88
Proudhon, Pierre-Joseph, 171
Puladin, 6, 248
Pushkin Square, 217

Q
Qaraje-Daghi, Hajji-Mulla Ahmad
    Mojtahed, 17
Qatran, 17, 28
Qavam-ol-Saltaneh, 5
Qazvini, Mohammad, 76, 86, 87, 93,
    143, 158
Quantum Mechanics, 59, 62, 183
Queen Jahan, 170, 178
Queen Victoria, 56

R
Radmanesh, Reza, 214, 225, 257, 258
Ra'i'yat, 19, 158
Rajabi, D., 209
Rakosi, Matyas, 258, 270
Rapallo Treaty, 186, 193, 218
Raskolnikov, Fiodor, 174
Rathenau, Walther, 44, 58, 65, 69, 82,
    165, 186
Ribot, Theodule-Armand, 106
Richards, Theodore William, 56
Riemann, Bernhard, 68, 98
Rivera, Diego, 45
Rolland, Romain, 45
Röntgen, Wilhelm, 58

Rosen, Frederic, 102
Rosen, Friedrich, 4, 31, 34, 42–44,
    79–84, 86–88, 93–98, 140, 141,
    187, 191, 212, 213, 220
Rosenberg, Alfred, 46, 185, 240
Rosenmund, Karl Wilhelm, 49
Roth, Joseph, viii, 47, 72
Rothstein, Theodore, 179, 189
Rubaiyat, 81, 83, 87
Rutherford, Ernest, 58

S
Saadi, 73, 81, 92, 99, 100, 102, 119,
    158, 161
Sachau, Eduard, 79
Safavid, 23, 32, 33, 170, 273
Sanaye'e Alman va Shargh, 50, 77,
    144
Saqafi, Khalil, 38, 106, 115
Sardar Asad, 6, 12, 15, 229, 262
Sassanid, 6, 100, 159, 168, 232
Sattar Khan, 22, 25, 26, 28, 74, 79
Schabinger von Schowingen, Karl
    Emil, 78
Schacht, Hjalmar, 4, 12, 157, 187,
    193, 237, 238, 255, 256
Schapschal, 24
Scharaf Mosaffari, 21, 34, 192
Schiller, Friedrich, 122, 141, 151, 201
von Schirach, Baldur, 13, 241, 256
Schlossplatz, 49
Schopenhauer, Arthur, 58
Schrödinger, Erwin, 62
Schtrüng, 4
Seif'e Azad, Abdol-Rahman, 77, 144,
    152
Seljuq, 31, 78, 92, 93, 102, 145
Seminar für Orientalische Sprachen,
    79, 88
Series in Exact Sciences, 43, 65, 69, 96,
    119, 123, 136, 183, 205, 208,
    211